T0093869

THE CURIOUS WORLD
OF SEAHORSES

Till Hein

Translated by
Renée von Paschen

THE CURIOUS
WORLD OF
Seahorses

The Life and
Lore of a
Marine Marvel

GREYSTONE BOOKS
Vancouver/Berkeley/London

Greystone Books Ltd.
greystonebooks.com

Cataloguing data available from Library and Archives Canada
ISBN 978-1-77164-988-9 (cloth)
ISBN 978-1-77164-989-6 (epub)

Editing for English edition by James Penco
Proofreading by Alison Strobel
Indexing by Cameron Duder
Scientific review by David Booth
Cover and text design by Fiona Siu
Cover images by Freshwater and Marine Image Bank (seahorse);
Elodie_M (background) / iStock.com
Expert review of illustrations by Miguel Correia

Though science writing typically uses the
metric system, imperial measurements have been used in
this book for ease of reading. This translation has been
adapted and abridged for an international audience.

Printed and bound in Canada on FSC® certified paper at Friesens.
The FSC® label means that materials used for
the product have been responsibly sourced.

Greystone Books thanks the Canada Council for the Arts,
the British Columbia Arts Council, the Province of British
Columbia through the Book Publishing Tax Credit, and the
Government of Canada for supporting our publishing activities.

Canada

BRITISH
COLUMBIA

BRITISH COLUMBIA
ARTS COUNCIL
An agency of the Province of British Columbia

Canada Council
for the Arts

Conseil des arts
du Canada

Greystone Books gratefully acknowledges the xʷməθkʷəy̓əm (Musqueam),
Skwxwú7mesh (Squamish), and səlilwətaɬ (Tsleil-Waututh) peoples on
whose land our Vancouver head office is located.

Contents

Prologue

THE GREAT INDIVIDUALISTS
OF THE SEAS

*"At the bottom of the ocean, tail curled
around seagrass, the male seahorse sways back
and forth in the current. He may be tiny and mysterious,
but no ocean creature compares to him."*

HILDE AND YLVA ØSTBY

WHEN THE BRITISH marine biologist Helen Scales saw a seahorse for the first time, while diving in Vietnam, she was enchanted. "Seeing one," she writes, "felt like glimpsing a unicorn trotting through my garden." For my part, I am more of a snorkeler than a diver. I had only been acquainted with seahorses from the aquarium in the zoo.

I have since learned that seahorses are the great individualists of the seas. Real freaks! Even marine biologists are amazed by the mere existence of these creatures. "When God created the seahorse," says marine biologist and fish expert Jorge Gomezjurado, who has been researching the animal for years, "he may have had one too many." It's possible that God went beyond his limit. In any case, he wasn't able to find an equally bizarre construction design for any other creature. A body with a pouch like a kangaroo, independently moving eyes like a chameleon, a long snout like an anteater, as well as a prehensile tail like a monkey—complete with a crown on its head as unique as a human fingerprint. What could all that be good for?

The life of a seahorse is just as strange as its physiology, as I was to discover while researching this book. Humanity could learn a lot from this little creature. The horses of the sea have certainly taken the adage "slow and steady wins the race" to heart. In fact, seahorses even hold a record for their extremely leisurely speed: according to the Guinness World Records, the dwarf seahorse (*H. zosterae*) is the slowest fish in the world. Even snails are first-class sprinters compared to most seahorses. It's questionable whether a seahorse would even be able to pass a beginners' swim class, on account of its slow tempo.

To some, their aura might seem meditative. However, seahorses are carnivores despite their cute appearance. Their enormous appetite is just as characteristic as their sluggish speed. At an age of two weeks, infant seahorses already consume up to four thousand miniature shrimp each day.

Seahorses are surprisingly loud—they don't whinny, but while they're hunting, flirting, or being confronted with anything frustrating, they emit unusual clicking and buzzing noises. The strange thing is that seahorses risk attracting the attention of predators due to all the noise they make. Researchers, such as the zoologist and bioacoustics expert Friedrich Ladich from the University of Vienna, are investigating why seahorses can't keep their mouths shut (see chapter 7), despite the danger.

In the science world, seahorses have been known as *Hippocampi* (singular *Hippocampus*) since the 1570s. The genus is abbreviated as *H.* in this book. It's followed by the Latin name of the species—for example, *H. zosterae* for dwarf seahorses. "Hippocampus" is the name of the sea monster of ancient Greek mythology. *Hippos* means "horse" in ancient Greek, and *kampos* means "sea monster." The head of that creature really does look like a horse, while its body or trunk resembles a fish or serpent.

Yet what are seahorses, from a biological point of view? In the medieval ages, European merchants thought they were baby dragons from distant islands. And early natural scientists classified seahorses as "insects of the seas." In fact, they belong to the realm of fish (specifically, the superclass Osteichthyes), although they might not seem to at first glance, lacking scales.

Seahorses are highly diverse. Some fully grown seahorses are smaller than a human fingernail, while others are up to fourteen inches long. Dwarf seahorses (*H. zosterae*) don't live any longer than one year, but long-snouted seahorses (*H. guttulatus*) have a life span of up to twelve years, at least in an aquarium. On the other hand, dwarf seahorses reach sexual maturity before the age of three months; thus three generations of the species can be born in a single year.

More than 120 different species of seahorses have been described by scientists over the centuries. However, some of these scientists were overenthusiastic. In fact, there are not even half as many species of these creatures, according to more recent research. Yet it's possible that some seahorse species have not yet been discovered, as these fish are masters of disguise. Many of them can change their colors according to their moods, from grayish blue to mossy green, or from purplish red with pink knobs to yellow with orange-colored bumps. Other species either have black stripes, yellow dots, or a green-gray camouflage pattern. Many researchers are convinced that the changing colors serve not only the purpose of camouflage, but also act as a method of communication with members of the same species, as do their clicking and buzzing noises.

Some great minds had a passion for the horses of the seas. "The wonders of God and the craft of mother nature are visible in many wonderful creatures," wrote the famous Swiss naturalist Conrad Gessner in the sixteenth century, "especially in this present sea creature or fish." Yet, like the most fascinating things in our world, seahorses can also polarize opinions. "I consider them boring and mindless creatures," wrote Alfred Brehm about them, in the famous German compendium *Brehm's Life of Animals*, around three hundred years later.

Whether seahorses have highly developed minds is debatable, as a matter of fact. Some zoologists assume that fish have something like a "consciousness" and can reflect upon themselves and their behavior. But perhaps seahorses and other finned creatures don't have to constantly mull over everything.

Like the singing humpback whales and orcas, seahorses are artists; their talents are primarily in the fields of expressive dance and color design. As seducers, they are particularly impressive. Their wedding dances would melt any salty dog's heart on the spot. To greet each other, the male and female nod gallantly at each other. Then they curl the tips of their tails together and flirt around while tightly entwined. Sometimes they stop for a moment and rub their snouts together like a kiss. They circle around each other repeatedly, letting their bodies shimmer in various colors. Their wedding dances can last for up to nine hours. And while people usually get married only once (if ever)—to the same person, at least—seahorses repeat their wedding dance every couple of weeks. On top of this, behavioral scientists have found that many seahorse couples remain loyal to each other till death. In the meantime, it has come to light that these fish also have other kinds of relationships. According to their circumstances, some seahorses may tend toward changing partners or group orgies.

They are neither holier-than-thou nor classic heroes; however, as creatures that make the best of their circumstances, they get around the globe surprisingly well—as long as humans don't pollute the seas where they live. Many people think seahorses are only at home in the warm waters near the equator. In fact, they live in many different regions of the seas. With a lot of luck and patience, you can find them almost anywhere in the world—other than in snow and ice. However, most seahorses do populate the coastal regions of the seas in tropical and temperate latitudes. There they can be found hiding in the seagrass or mangroves. Others prefer coral reefs, or the mouths of rivers or estuaries. But even far out at high sea, it's not rare for fishers to reel up seahorses from the depths of the ocean as bycatch in their dragnets.

For over a couple of thousand years, seahorses have been inspiring our imaginations. In Greek mythology, it is seahorses that pull the sea god Poseidon's chariot, with mermaids riding on their backs. Some cultural historians also hypothesize that the figure of the knight in chess (in the shape of a horse head) was not modeled after a horse, but a seahorse. In Europe, on the other hand, the fictional "baron of lies," Baron Munchausen, claimed in the eighteenth century that he rode on the back of a seahorse. In the popular Pokémon video games from Japan, a killer seahorse-creature creates havoc. The names of real seahorses might even inspire our dreams: zebra seahorse, paradoxical seahorse, tiger tail seahorse.

Some people love seahorses more than anything else. For example, Elena Theys (see chapter 1) gave up her profession as a banker many years ago to dedicate her life to the amazing horses of the sea.

In the incredible world of seahorses, the males become pregnant—unique among nearly all the creatures of the world. How and why this came about is a mystery that researchers from various countries are currently trying to solve. Gender theorists are fascinated by the strange occurrence of male pregnancy, providing them with a model to further question traditional gender roles in human society.

Seahorses are so special that healing powers have been attributed to them throughout the centuries. During the Renaissance era in central Europe, they were used as a miraculous cure for poor vision and side pains, as well as rabies and low libido. "These creatures are said to induce a person to become unchaste," wrote Conrad Gessner. "Item dried, powdered and ingested, should provide a wonderful antidote for those who have been bitten by mad dogs." Until the eighteenth century, seahorse remedies were found in many home apothecaries, where they were used to alleviate exhaustion, hair loss, or impotence, among other ailments.

In Asia, doctors still prescribe large quantities of seahorse antidotes. A classic remedy in traditional Chinese medicine (TCM), which ostensibly helps alleviate ailments of almost any kind, is pulverized seahorse mixed with honey, ginseng, and red olives. However, seahorses are primarily prized in TCM as a sort of natural Viagra, which has also spurred the curiosity of some men in the West. A seller of bogus impotence remedies in Germany offers his products under the name Super Hard, as antidotes for low libido, erectile dysfunction, and premature ejaculation. Along with ginseng and stag's penis, the product contains dried ground seahorse — which means the pulverized body of the only male creature in the world to play a "female" role in reproduction is thought to help strengthen the male member!

Yet people haven't only looked to seahorses for help if they were plagued with libido problems. For thousands of years, these creatures have inspired mystics, painters, artists, and storytellers around the world. Jewelry, carvings, murals, and coats of arms are decorated with their images, as well as everything from antique vases to modern toilet bowls. Seahorses served as good luck charms to ward off evil spirits and were supposed to bring good fortune to sailors and fishers.

Scientists, designers, and engineers are placing their bets on seahorse power. At the end of the twentieth century, a Japanese research team found that seahorse-shaped pillows facilitate highly rejuvenating slumber. The nifty design of the seahorse's hindquarters serves in robotics as a model for robust and flexible prehensile devices. Yet how does the seahorse itself benefit from its unusual body structure? Are seahorses related to horses or hippopotamuses? How many species of seahorses are there? How did they evolve? What are their daily routines? Can we learn more about the essence of masculinity from these fish? Is there any truth in the supposed healing powers of dried seahorses? Are they suitable pets for people with a day job? Did Indigenous Australians revere the seahorse, rather than the Rainbow Serpent, as their major creator spirit? Could genetic research on these fish help people with HIV/AIDS? And can seahorses, whose greatest enemy is *Homo sapiens*, be protected from extinction?

Experts are increasingly worried about several species of the seahorse genus. Dragnet fishing—or "bottom trawling"—and its destruction of habitats has particularly harsh consequences for seahorses. Some animal rights societies and environmental protection groups hope that the seahorses' beauty and charisma will bring humankind to its senses, thereby making a decisive contribution to saving the seas in the form of a magical mascot.

The Curious World of Seahorses is an homage to these great individualists of the seas, which prove that everything in the world is truly astonishing—including normality.

Yellow Seahorse

Hippocampus kuda

Seahorse Whisperer

THE AQUARIUM KEEPER IN GERMANY

―――――――

"Being a geek is a great thing . . .
[It] means you're passionate about
something and that defines your uniqueness.
I would rather be passionate about something
than be apathetic about everything."

MASI OKA

A LIGHT-GRAY SKY hangs low over the flatlands. Rain washes over the roads. Somewhere on the main street, flanked by apartment buildings, fields, and huge puddles, a sign on a little house reads "Seahorse Pet Shop, 24." Here in the middle of nowhere, in the German town of Visselhövede, the seahorse whisperer, Elena Theys, works.

Soon I'm standing beside her in astonishment in front of a large aquarium. Six white and brown speckled seahorses have wildly entwined their prehensile tails. As though in a trance, they are cuddling together and pressing their bodies up against the glass plate. I've read that seahorses are highly social creatures—and they love cuddling together. During courtship, the male and female often entwine their hindquarters and promenade through the underwater world holding each other's tails. But what's going on here? An orgy?

Theys shakes her head. "The seahorses are simply following their instinct and holding on to something," she explains and laughs. Lacking any other convenient grips, they use their companions—that's the explanation for their entwined group cuddle. In contrast to seahorses in the wild, these creatures are used to people, the pet shop owner emphasizes. "And at the moment, they're totally focused on us." In fact, the seahorses look like they're staring at us, just as we're staring at them. Theys whispers, "They're trying to find a way to get to us."

Theys notes that seahorses are highly sensitive. They have "very diverse" reactions to different people—and the creatures in the aquarium have apparently taken a liking to me.

Elena Theys, fifty-nine, can be taken at her word. She has been keeping fish in aquariums since 1986. Over two decades ago she set up her first saltwater aquarium, and in 2004 she fell in love with seahorses. With her rock-star mane of hair, her tie-dyed sweatshirt, and her joggers, she looks like a teacher from an alternative kindergarten. However, before she dove into the deep end and became a seahorse whisperer, she was a banker. Theys has been selling seahorses and other marine creatures to enthusiasts from all over Germany,

Switzerland, and Austria for more than ten years. Many of her customers buy the creatures right in her shop. Others have them shipped by courier, in plastic bags filled with sufficient saltwater and packed in between polystyrene sheets. Her least expensive fish, young guppies, cost around 75 cents each; a cleaner shrimp can be had for the equivalent of 15 US dollars. Seahorses are for those with deeper pockets. Such a creature costs around 100 US dollars. Many of her customers want to have a pair; Theys gets around 250 US dollars for a couple.

Without this woman's expert advice, more than a few professional aquarium operators in zoos would have been stranded. Whether it's a question of reproduction, nutrition, or disease, if the professional experts are at a loss, the self-educated shop owner from this small town can often find a solution. Once her colleagues from the aquarium in Münster called in despair, asking for her advice. Their seahorses looked all lumpy. Gas bubbles had formed under their skin. "When the aquarium water is oversaturated with oxygen and nitrogen, this dangerous condition can arise," explained Theys. "Many seahorses die because of it."

As recommended in professional sources, the people in Münster immediately transferred the sick seahorses to a quarantine pond. When the symptoms disappeared, they put the seahorses back into their old aquarium—and once more, blisters immediately formed under their skin. There isn't any medication available to treat gas-bubble disease. What to do? Theys recommended that they reduce the aquarium temperature to 73 degrees Fahrenheit and stop the activated carbon filtering. The seahorses immediately recovered.

Theys hasn't spent a single day without seahorses for the past sixteen years. She's constantly learning something new about her fosterlings. "They turn away from some people immediately," she says, while we continue watching the entwined white and brown seahorses in her aquarium. Recently, for example, a girl dragged her grandmother into the shop. The girl was obsessed with seahorses, but her grandmother wasn't interested in the least. "As soon as the elderly lady approached the aquarium, the seahorses hid away in the farthest corner," Theys recalls. "And you can see for yourself," she adds, "they feel comfortable now."

The six entwined creatures belong to the species known as lined seahorses (*H. erectus*), a species at home in the western Atlantic, Theys says as she gently runs her index finger up and down the glass pane of the aquarium. They probably acquired their colloquial name from their narrow horizontal stripes.

Suddenly, they look as though someone had painted them with white varnish. Has the lighting in the shop changed? Or have the creatures really turned paler? "Quite impressive, aren't they?" asks Theys and smiles. "Whenever I pet them, they turn lighter." The seahorses seem to like affection— glass pane or not. "During courtship, lined seahorses always turn paler too." Theys considers this proof that the ability of many seahorses to change color serves not only the purpose of camouflage, but also of expressing their feelings.

Biologists concur with this theory as well. However, this self-educated woman has an ambivalent relationship with science. "More marine biologists have benefited from us aquarium keepers than vice versa," states Theys confidently. Some researchers have even propagated "myths" about her favorite creatures. "I strictly prohibit my seahorses from reading the professional literature." More than a few professionals recommend that seahorses should never be kept with other fish, Theys complains. "But that's total rubbish! All my experience over the past sixteen years goes to prove the opposite."

She divided her first aquarium into two compartments using gauze—one for the seahorses and the other for clownfish and surgeonfish. However, the material soon broke loose from its attachments and the artificial separation was breached. "The only ones that were really irritated were the clownfish," Theys remembers. "They needed a week to get used to their strange, new, scaleless neighbors." The seahorses, on the other hand, were neither afraid nor shy—they got along really well with the other fish. Since then, Theys has been convinced of the benefits of multicultural life in an aquarium. She has also successfully kept seahorses in the same aquarium as dwarf angelfish and dwarf wrasses. Yet keeping them together with cleaner shrimp can be risky: some of the crustaceans might get hungry enough to devour the new generation of seahorses developing in a male seahorse's brood pouch before the babies are born. Theys is nonetheless convinced that seahorses generally prefer living together with many different species of fish, rather than alone with their own kind, despite what scientists say.

Theys regularly holds guest lectures for aquarium keepers all over the country. Her expertise is also available via You-Tube. One of her messages is: "If seahorses in an aquarium are starving, don't ever blame the other fish." Most other finned creatures are much faster than Theys's favorites.

Then what could be the problem? "Some people have two dogs, for example, yet one of them may be dominant," she says. "You'll have to give them both enough dog food, so there's still plenty left over for the weaker dog after the dominant one has finished feeding. The same thing applies to fish."

If you believe internet sources, these creatures are said to require very little space. On websites of seahorse aficionados, it's said that a twenty-one-gallon aquarium is sufficient for three pairs of seahorses. "Sure thing," scoffs Theys, grimacing. "Big-belly seahorses [*H. abdominalis*] can grow to a size of fourteen inches," she says. "You'd have to stack them on top of each other in a tiny aquarium like that."

There are dozens of seahorse species in the world, she explains, and each of them has its own distinct characteristics. Theys follows several rules of thumb. She recommends experienced saltwater aquarium keepers get a tank that holds at least thirty-four gallons for two to three pairs of seahorses. However, newcomers should use a tank almost double that size for the same number of creatures, since it's not always easy to keep an eye on everything due to the large amount of feed required. The seahorses' bony plates make it difficult for inexperienced aquarium keepers to recognize whether their abdomen are sunken due to malnourishment.

Her magic nutritional formula is that each seahorse needs around five times as much feed as another fish of the same size. No matter whether it's a dwarf seahorse (*H. zosterae*), which isn't quite an inch long, or a huge big-belly seahorse (*H. abdominalis*). "Seahorses don't have a stomach, only intestines," explains Theys. That means they don't have a good feed conversion ratio. "This really proves what excellent hunters they are. Otherwise, evolution would definitely have given them a better digestive system." Another rule of thumb: you need to put sufficient feed in the tank so that the seahorses can eat their fill for at least sixty minutes, without other, hungrier fishes competing for their feed. Then you're on the safe side. Apart from that, she encourages seahorse keepers to be courageous enough to experiment. "Life is an adventure," she says. Varying impulses and stimuli are usually good for horses of the sea.

It's lunchtime! Theys opens her large cooler and takes out a box. Then she lifts the aquarium lid and gently drops a snow-white cube into the water, a bit larger than a sugar lump for coffee. The cube floats in the tank for a while, then it dissipates. Innumerable bits of fuzz begin forming like snowflakes. "Dwarf opossum shrimp," explains Theys. "They were deep-frozen and pressed into a cube, and now they're thawing in the lukewarm water."

Leisurely, in an upright position, two seahorses approach the feed, just like in nature documentaries. Without any hurry, they stalk the fuzzy dwarf shrimp. And all of a sudden—gulp! The shrimp all disappear in an instant. Had I stopped paying attention for a moment? The marine life specialist shakes her head. "Seahorses suck up their prey through their tubular snout faster than the human eye can perceive." They particularly like mysis shrimp. When there's nothing better to be had, they'll also feed on brine shrimp (*Artemia*). "Yet if newly purchased seahorses still aren't feeding after three

17

days, it's best to get expert advice from an aquarium pet shop or a veterinarian."

Theys's passion for seahorses began with a misunderstanding. In 2004, she was visiting her old hometown of Augsburg in southern Germany for the first time in a while. In passing, she took a brief look at the display window of the local pet shop. The display was decorated with two huge, luminescent, orange seahorses. "Wow," thought Theys, "what wonderful lamps!" But the creatures suddenly began to move. "Those aren't lamps at all," she gasped. "They're live seahorses in an aquarium!"

Theys couldn't get enough of the sight. She rushed into the store and asked the salesperson an endless barrage of questions. The two horses of the sea, almost twelve inches long, had been caught in the wild in Brazil—and they were a mating pair. She wanted to know how the salesperson could be sure. In the Augsburg pet shop, she heard about the wedding dances and male pregnancies of these fish for the first time.

She bought the two finned creatures for the equivalent of just under 80 US dollars. She took the pair of seahorses safely back home to northern Germany in a plastic bag filled with saltwater and put them in an aquarium—and that's how her unusual story began.

Normally, she doesn't give her seahorses names, Theys tells me, flicking her long hair off her forehead. "I often have enough trouble handing over my seahorses to customers." Names would only increase her separation anxiety. However, if certain seahorses have really found their way into her heart, she no longer follows her own rule. She'll never forget the seahorse Charlie, the strongest male from her herd of "Brazilian giants." An orange-colored figurehead decorates the entrance door to her pet shop to this day. "Charlie lived for over seven years," recalls Theys, as her whole face beams. For a seahorse, that's a biblical age—but let's not get ahead of ourselves.

With wild-caught seahorses, the pet shop owner in Augsburg warned her in 2004, nutrition is the key. Those seahorses will only accept live feed. Otherwise, they'll go on a hunger strike. After brainstorming, Theys got in her car and raced to the North Sea at Germany's coast, where she sweet-talked the shrimp fishers till they gave her the bycatch from their nets. Two huge pails were filled with tiny opossum shrimp, scuds (Amphipoda), and sea lice. Her seahorses, at home in their tank, were highly enthusiastic. The pair from Brazil hungrily gulped up their treats. Since the episode with the gauze separation wall, her favorite creatures have shared a tank with clownfish and surgeonfish. Her Brazilian giants seemed to be happy as punch with live feed from the North Sea. "Soon they began dancing together," Theys recalls.

The paperwork, on the other hand, proved more complicated. Many species of seahorses were considered threatened. Therefore, as Theys discovered, purchases had to be registered with the local nature protection authorities. Otherwise, she would be risking a fine. Yet, to register the creatures, she had to know which species they were.

She was already aware that there are many different species of seahorses. And her pair was caught in the wild off the coast of Brazil. But which species? The former banker flipped through the professional literature, including hundreds of scientific descriptions—but she couldn't find her seahorses anywhere. Eventually, she gave up. Officially, there was no mention of her species of seahorse; hers were probably very rare creatures.

"By that time at the latest, I had realized that I wanted to become a seahorse breeder," Theys says. However, she needed more space for her undertaking. Her living room was unsuitable. She set up her first breeding aquarium in her laundry room. After a few months, it became clear that she wouldn't be able to handle the work without a helper. But at least the giant seahorses from Brazil quickly got used to frozen feed—and a few months later, they had already given birth to a new generation. "Those beautiful, dark yellow seahorse babies grew so fast. That strengthened my resolve to never give up, despite all the setbacks that were soon to come," says Theys.

Her laundry room also quickly became too small, so Theys moved her seventeen aquariums into a garage. Together with like-minded friends, she founded a club for seahorse aficionados and gave tours of her seahorse zoo. A TV team even shot a nature documentary in her garage. She had more and more work. She hired long-term unemployed people to help her. "Some of them didn't return after the first couple of days," she says. "It was too stressful for them." But others were able to cope with the work and found a steady job. "You see," Theys says, beaming with satisfaction, "seahorses can also help people do good."

The shopkeeper asks me to excuse her for a while and seats herself at the desk behind her counter, which is decorated with seahorse memorabilia. She needs to take care of a few orders. In the meantime, I can watch the slender seahorses (*H. reidi*) in the tank on the far left-hand side.

Yet where are they hiding? I can see water, green seaweed, pieces of stone, and a few little, gray fish. Not a seahorse in sight. Finally, I do spy one. Beige with gray speckles, it's slinking along the back of a stone. Then I see another. It's more corpulent and is hiding behind another stone. Its armor of bony plates makes it look just like it's made of rock. The perfect camouflage! But why are they playing hide-and-seek? The group of entwined creatures this morning was so trusting! Do these seahorses not like me as much as the others did?

"It's probably just a misunderstanding," I think to myself after a while. Memories of a family holiday in Italy flit through my mind. During the first week, I thought the Italians found my broken phrases charming and were being kind to me for trying to speak their language. Until one morning, when I went to the bakery without my four-year-old son— and no one smiled at me anymore. The lesson I learned was that it's not always just you who sets the mood. Your company can also make a world of difference.

In general, seahorses react to me very enthusiastically—as long as Theys is standing next to me. As soon as the seahorse whisperer joins me in front of the tank, the pair of four-inch-long scaredy-cats with their tiny pipette-like snouts and ringlet tails peek out from behind their stones. "The scientific name of the slender seahorses is *Hippocampus reidi*, and their habitat is the western Atlantic," the marine aquarium expert tells me. They seem like strange creatures.

The thinner seahorse is not swimming upright in a classic seahorse position. With its tail unfurled, it's creeping along the floor of the aquarium like a snake. "Seahorses are highly versatile and really unique fishes," says Theys. Their gills, for example, are completely closed, other than two tiny holes on the left and right of their necks. And the appendages that look like ears are, in fact, pectoral fins.

"Is that fat guy on the right pregnant?" I ask. The pet shop owner laughs. At least I seem to know that the females don't get pregnant. "The fat man isn't pregnant," says Theys. His brood pouch has just become enlarged, because he's in love. In the case of *Homo sapiens,* men often like to boast with their fat wallets. But in the realm of seahorses, the males make a big splash with their large pouches to impress their chosen partner.

The shop door rings, and Theys needs to serve her customers. Toward the end of the working day, she finds time to talk about her favorite subject again, the Brazilian giants that marked the beginning of her passion for seahorses. In 2005, a biologist named Tim Migawski from the University of Oldenburg conducted his research in Theys's garage instead of on-site in the sea. Several times a week, he weighed her South American seahorses and counted their offspring. He calculated their reproductive cycles, length-to-weight ratios, specific growth rates, as well as the frequency of pregnancy and length of gestation. After three years, in 2008, he successfully submitted his graduate thesis, titled "Parameters of the Life Cycle of Brazilian Giant Seahorses."

A huge number of seahorse offspring were born in aquariums in Theys's garage back then, she tells me with shining eyes. Around 34,000 all together. "And I managed to raise several hundred of them." She's proud of dedicating her life to seahorses. "But I often paid a price for learning the hard way." The first baby Brazilian giants all died within the first three days. The second brood died after three weeks. When the male gave birth to offspring for the third time, Theys kept him in a special tank to be on the safe side. Promptly, everything seemed to run smoothly. Yet, after three months, when they were introduced to frozen feed, those juvenile seahorses died as well.

Large broods of seahorse babies were born over and over. The largest brood of babies totaled 1,472 infants. Theys counted them repeatedly. And when she fed them tiny zooplankton, a new generation finally matured. Several hundred Brazilian giant "sea foals" grew up, each up to two inches long. Everything was going well until one night, when tiny jellyfish appeared in the aquarium. They quickly grew from one millimeter in size to one inch, gripping the backs of the seahorses like rodeo riders. Quite soon, white spots formed on the seahorses' skin. They began to have trouble breathing, and many died. Had the jellyfish deployed their poisonous stingers?

Theys and her helpers removed the survivors and put them in another tank. They emptied the original tanks and cleaned them thoroughly. But as soon as the seahorses were reintroduced to their old domicile, the same problems recurred. Somehow, the jellyfish had survived unnoticed in the old tank—and they reproduced at high speed once more. Experts advised her to use a special chemical treatment to kill any parasites. The chemicals were said to be harmless for fish. However, the experts had fish like carp or goldfish in mind. Seahorses have more sensitive gills than most other fish. The Brazilian giants' respiratory organs were damaged by the chemicals. All of the adult seahorses died.

Almost miraculously, Theys was able to transfer eighty-one of the baby seahorses into another tank before it was too late. Several dozen developed well and were still alive by the summer of 2008, having grown up to twelve inches long. However, the seahorse whisperer discovered that their magical orange coloring, which she had liked so much, was fading progressively with each new generation. Formerly, the creatures had often changed their color from black to dark yellow—but at some point in time, they remained black.

Theys sold the last of her Brazilian giants over ten years ago. She isn't sure whether any aquarium keepers still have offspring from the seahorses she bred. "Maybe someday a former customer will get in touch," she says. "I'd be pleased!"

Whether the Brazilian giants really are a newly discovered species remains unclear to this day. The seahorse expert Rudie H. Kuiter praises Theys's breeding success in one of his specialist publications. He belongs to the fraction that's convinced the Brazilian giants are a unique species and not "only" a particularly pretty subspecies of the slender seahorse (*H. reidi*). In his expert handbook *Seahorses and Their Relatives*, Kuiter gives them the provisional name of *Hippocampus c.f. reidi*—"c.f." meaning "comparable to." He hopes this species will soon be recognized by scientists and given a suitable name.

Theys takes down a thick binder from a shelf full of specialist publications and the account books for her shop. She flips through newspaper reports about her work. Is it true that seahorses are threatened with extinction? The pet shopkeeper stops for a moment to reflect. "Yes and no," she eventually answers. Some species are still widespread in many seas to this day. She knows this from her contacts with seahorse dealers in different parts of the world. "A few species don't need any special protection." The problem is that no layperson can distinguish the yellow seahorse (*H. kuda*), also known as the common or spotted seahorse, from similar but much rarer species. Especially if they have been dried, which is how these creatures are sold in huge volumes for the preparation of traditional Chinese medicine.

The market for seahorses as pet fish is much smaller. However, Theys emphasizes that catching seahorses in the wild for aquariums has long been prohibited. CITES, the Convention on International Trade in Endangered Species of Wild Fauna and Flora, which has been signed by 184 nations (see chapter 16), stipulates that only seahorses over four inches in length can be sold on the international market. This ensures that only around a dozen seahorse species are legally available on the market.

Theys emphasizes that CITES is fundamentally important. However, the devil lies in the details. For example, its special regulation regarding "personal or household effects" is peculiar. It stipulates that any international traveler can import four dead seahorses via customs worldwide. These are considered "personal effects," and customs officers are usually accommodating. "However, if you have even a single live seahorse in your luggage, you'll get into very serious trouble." Confiscation is the least of all the potential problems. "Totally absurd," says Theys. "Politicians urgently need to revise these regulations."

For many years, Theys cared for two seahorses from the North Sea on a daily basis. She pulls a yellowed photo from a folder with newspaper articles, letters, and other memorabilia. "This is where Fritzchen from Cuxhaven lives with his wife Frieda," it says in the caption below a photo of a tank. The picture was taken at a permanent exhibition in her Seahorse Club, she says. "Those were good times."

In August 2008, a fisher from Cuxhaven on Germany's coast caught one of the rare North Sea seahorses in his net—a short-snouted seahorse (*H. hippocampus*). The fisher brought the creature in a plastic pail to the large Sea Life aquarium in Cuxhaven, where he was named "Fritzchen" and kept in a small tank with the bare necessities and no companions. "His favorite hangout was a dark green water plant made of plastic," recalls Theys.

Local journalists almost made as much fuss about Fritzchen as they did when the baby polar bear Knut was born a few years later in the Berlin Zoo. However, Sea Life soon had to close its doors for budgetary reasons. What to do with Fritzchen? The Seahorse Club offered asylum to Fritzchen, and the nature protection authorities gave the go-ahead. Fritzchen became Theys's fosterling.

In October 2009, he was moved into a large aquarium. His favorite plastic seaweed moved with him. Fritzchen clung on to it like a little child holding on to his favorite stuffed animal, even when Theys had to take the plant out of the water to clean it. "Fritzchen was clinging on as though his life depended on it," she recalls. "Only after a while did he save his own life by bravely jumping back into the water in his tank. Afterward, he looked completely lost, until I gave him back the plastic water plant." Plastic algae or not— seahorses can only survive on land for a short time. Their gills only enable them to breathe underwater.

One after another, different kinds of creatures joined Fritzchen in his tank—sole, eels, starfish. "That gave him all sorts of new stimulation," says Theys, suddenly sounding like a social worker. He often waved his fins around wildly to get their attention. "And many children and adults learned from Fritzchen for the first time that seahorses are also at home in the North Sea."

The best thing that happened to him was the day he met Frieda. The female seahorse came to the Seahorse Club from Kiel. "She was born in an aquarium and had never seen another fish other than her parents and siblings," says Theys. When Frieda joined the party in Fritzchen's tank, all the current residents circled around her to check her out. That made her really shy. "But when Fritzchen saw Frieda, he was immediately infatuated," Theys recalls. Finally, a female on the block!

Three days later, Frieda had overcome her fear of the other residents, although many of them were much larger than her. Fritzchen courted her—and they became inseparable. "They often cuddled together and dozed off closely entwined in the branches of the plastic plant." Frieda and Fritzchen danced together every day, changing color—typical for this seahorse species during a display of positive emotions—lighter bellies, backs, and tails, and darker faces, Theys says. Unfortunately, they were not able to have any offspring.

The seahorse whisperer pauses for a moment, lost in thought. "A saltwater aquarium is a little like a garden," she says. You don't need any highly specialized know-how to manage it nowadays. "Nevertheless, it's a biotope that requires personal responsibility and good care."

When a large brood of baby seahorses is born in an aquarium, many of the infants often die, although Theys once managed to successfully raise over three hundred babies of a rare seahorse species. "But I had to get up every two hours each night for four months to feed them," she recalls. "I don't know if I could handle that again."

Whenever seahorse babies are born in her marine pet shop, she's as gleeful as a little child. However, she's no longer been actively breeding seahorses for several years, in part because it's too much of a financial burden. At large breeding stations with electronic feeding machines in warmer climates, such as in Asia, the situation is quite different. But by caring for everything by hand in the northern German climate, with small numbers of seahorses, she'd often invested up to the equivalent of 400 US dollars in a giant seahorse, which she was only able to sell for 100 US dollars. The feed is particularly expensive, in addition to the rent and heating costs. Yet sometimes she recalls her former enthusiasm; perhaps she'll start breeding seahorses again one day.

Another option would be taking care of the rare species confiscated by customs officers, and using them for breeding purposes. "Those seahorses need to be taken care of somewhere," she says in a fighter spirit. "And I would receive CITES papers for their offspring, since they'd be aquarium-born." Another advantage is that aficionados are more willing to pay reasonable prices for rare species.

However, nature is unpredictable, especially with seahorses. "Once I raised 300 babies of a sought-after species," she remembers. A huge success. "Yet, there were only 30 females in comparison to 270 males. And almost all customers want pairs." She ended up with around 200 male seahorses that she wasn't able to sell, so she took care of them until the last one died at an age of seven years old.

Why one gender clearly prevails in almost every brood remains a source of controversy among seahorse researchers. Perhaps the water temperature plays a role in the development of biological gender—that seems to be the case with certain fish species. The aromatase enzyme is probably an important factor, since it can transform male sex organs into female sex organs. Scientists have determined that aromatase reacts to changes in temperature—and so might other enzymes that promote masculinization. Recent studies have also demonstrated that warm water promotes the development of male sex organs in fish. A study of fifty-nine fish species has shown that increasing the water temperature by one to two degrees can alter the ratio of female to male from 1:1 to 1:3. "The future is female" is a common slogan for combating the human patriarchy. On the other hand—in view of global warming—it looks like the future of many fish species might be male instead.

Does this tendency also affect seahorses? Theys shrugs her shoulders. "As juveniles, they all still look female." But by the age of six to twelve months, a part of the male sexual organs has developed—in particular, the brood pouch. However, there isn't any rule without an exception. It seems that seahorse gender may also be interchangeable in later years. A female in the marine pet shop suddenly changed into a male at an age of two years—after it had already produced offspring several times via its eggs.

Anyone keeping seahorses, Theys says with a smile, should always be prepared for a surprise.

= 2 =

Rainbow
Creatures

WHY TAIL FINS ARE OVERRATED

═══════

*"Peer at a seahorse, briefly hold one up
to the light, and you will see a most unlikely creature,
something that you would hardly believe was real
were it not lying there in the palm of your hand."*

HELEN SCALES

THE LION, as any child knows, is the king of the animals, even though it doesn't have a crown. However, almost all of the much tinier and more delicate seahorses have crowns. The physical features of these creatures make them quite a curiosity. Their torso has a pouch like a kangaroo, their eyes can move separately like those of a chameleon, and their long snout is akin to a pipette. They have strong tails for gripping holdfasts—and they also have majestic crowns. How did nature manage to come up with such an unusual composition?

31

The visual similarity between seahorses and land-dwelling horses is obvious. However, from their heads downward, the two have nothing further in common. The seahorse's body appears fragile, and its tail looks somewhat like that of a serpent. What about hooves? Nothing doing. These creatures seem like they've been put together from several completely different animals—and they inspired human fantasy to create remarkable chimeras in former times, such as the famous hippocampus of ancient Greece (see chapter 5).

Biologists count seahorses among the fish, although they have neither tail fins, nor ventral fins, nor scales. Instead, their body is protected by bony plates. So, what's fishy about them? Take a closer look and you'll find they're very fishlike. For example, seahorses breathe via gills, and they have a swim bladder—as well as several fins, although the organs they use to propel, steer, and stabilize themselves are minuscule. Their swim bladder serves to counterbalance the weight of their body armor, made of bony plates. It can release and absorb gas, thus regulating its buoyancy—typical of many fishes. Most of the seahorse's remaining bodily organs are like those of other finned creatures—for example, their heart, liver, and kidneys.

In general, whatever looks strange or bizarre about a seahorse has, in fact, a functional purpose. It's a question of the best possible adaptation to their environment, as well as their behavioral patterns. The most famous example is their highly flexible prehensile tail, which is self-rejuvenating. It replaces the tail fin, which most fishes flick sideways to propel themselves forward. Instead of using their lower bodies to move forward, seahorses use them to hang around, in the true sense of the expression.

The seahorse's tail is so long and flexible that they would be able to scratch their heads with it or use it as a scarf. However, the tail has a different function: it enables them to hold on to either finger coral or stalks of seagrass. That's why seahorses aren't at the mercy of the waves and currents in the sea—despite their tiny fins—and they can recover from their strenuous swimming. Were it not for their prehensile tails, they could even be washed ashore by the waves.

"Holding on is what these creatures do best," says marine biologist Ralf Schneider from the GEOMAR Helmholtz Centre for Ocean Research Kiel. "Seahorses aren't good swimmers; they're neither fast nor tireless." They'd never be able to catch any fast prey that way. Instead, they rely on camouflage and surprise attacks. And for a creature that doesn't want to be devoured while it's hunting, clinging on to something is a better strategy than floating in place. "However, there are other fish, such as gobies and blennies, that just rest on the floor of the sea as they recover," says Schneider. Saving energy can't be the only reason why seahorses attach themselves to something so often.

Instead, Schneider assumes it must be a strategy for avoiding predators. "It's very difficult to detach a seahorse that's clinging on to something tightly with its tail," says the marine biologist. And he should know, as that sort of finger muscle training is part of his daily routine during the experiments at GEOMAR in Kiel. "Many predator fish are not able to detach a seahorse with their jaws when it's gripping something tightly," says Schneider. This creature's prehensile tail is so strong and cleverly constructed that it has inspired robotics researchers to design prosthetics for human beings (see chapter 14).

Seahorses do not have stomachs. Their nourishment goes directly from the esophagus to a foregut. Their digestive tract only takes up 40 percent of their body length. (In the case of other carnivores, it's almost 100 percent, and many herbivores in the animal kingdom have intestines and stomachs that are three times as long as the length of their body.) "Part of the nutrition is absorbed by specialized cells in the walls of the seahorses' intestines, and whatever can't easily be digested passes through the intestines and is quickly flushed out via the anus. Other waste products are filtered from the blood by the kidneys—oblong, paired organs, which, like those in human beings, are located on either side of the spinal column. From there, the urine flows to a sort of bladder, to be passed out via the so-called urogenital papilla," explains the seahorse expert and marine biologist Daniel Abed-Navandi, deputy director at the House of the Sea aquarium in Vienna.

The seahorse's simple digestive tract is a stress factor, since it means that these creatures are poor at utilizing nutrition. They need to feed almost constantly, and the industrially processed "dead" feed that's often used in aquariums can easily overtax them, because seahorses lack the digestive enzymes that other creatures produce via their stomachs.

Characteristic for these unusual fish, they only have pectoral fins, dorsal fins, and anal fins. The pectoral and dorsal fins can be flattened against the seahorse's body, which helps it maneuver in dense vegetation. This enables the horses of the sea to weave their way through life with ease. All seahorse fins are tiny, delicate, and transparent. The most conspicuous of these fins is the dorsal fin on the back, which is responsible for propelling the seahorse forward. The two pectoral fins, left and right of the neck, help it navigate, supported by the tiny anal fin—however, they create very little forward propulsion.

Some seahorses can flick their fins back and forth up to fifty times a second, as determined by scientists. These wavy motions are called undulations.

The newborn babies of several seahorse species also have a tail fin, in addition to pectoral, dorsal, and anal fins; however, the tail fin falls off after a couple of days. "A remnant of evolutionary history," explains Axel Meyer, professor of biology at the University of Konstanz. "The predecessors of all seahorses had tail fins." Meyer was able to partially clarify how the unusual physical features of the seahorse came into being millions of years ago. "We tried to find an explanation for seahorses' unusual appearance and behavior in their genome," says the scientist. Along with other scientists from Germany, China, and Singapore, Meyer decoded the DNA of the tiger tail seahorse (*H. comes*) and published the results in 2016 in the scientific journal *Nature*. Among other influences, the researchers determined that the special features of the seahorse's morphology are based on three factors—loss of genes, failure of so-called regulatory elements, and genetic recombination.

Scientists counted exactly 23,458 genes in the tiger tail seahorse's DNA. A very high figure. (Even human DNA doesn't comprise a higher number of genes. However, experts assume that genetic connectivity is more complex in *Homo sapiens* than in seahorses.) But what's the purpose of all the genetic material? And how is it regulated?

In fish, a gene named Tbx4 plays an important role in the development of pectoral fins, as was already known from other studies. However, this specific gene was missing in seahorses—exactly like the pelvic fins. Was this a case of cause and effect? Meyer and his colleagues tested this hypothesis by turning off the Tbx4 gene in zebra fish using CRISPR, a bioengineering tool used like genetic scissors. "Bingo!" exclaimed the evolutionary scientist. "These genetically manipulated fish also lost their pelvic fins—just like seahorses."

The horses of the sea don't need any pelvic fins, according to Meyer. While other predatory fish, such as sailfish and swordfish, can race though the water at top speeds of around sixty miles per hour, seahorses are much more sedate. And lacking any pelvic fins, they are less affected by the surf, the biologist surmises.

When it comes to swimming, seahorses have little in common with other finned creatures, since they assume an upright position in the water and their propulsion is anything but streamlined. The seahorse counts among the slowest fishes in the world. Dwarf seahorses (*H. zosterae*), found along the coast of Florida, Texas, Mexico, and the Bahamas, reach a top speed of only five feet per hour. In comparison, even land snails are twice as fast.

However, seahorses really have no reason to be in a hurry. Instead, they rely on good camouflage. Hidden from curious eyes, they lurk around waiting for prey. Many horses of the sea can change color to match their environment, and they happen to be talented disguise artists. They can take on the luminous orange color and the surface structure of a certain soft coral, or mime a yellow mangrove leaf floating in the water. Several species also have unusual skin flaps on their bodies that look like the leaves of water plants. In the tangles of an algae forest, seahorses can barely be discerned. Although the seahorse's miniature fins don't enable it to move far quickly, those delicate organs fulfill an important purpose: they make the creature agile and able to swim in very close proximity to dense coral reefs and water plants.

Why don't seahorses have any teeth, if they're predators? In the nineteenth century, advertisements appeared for "seahorse teeth" in British newspapers. However, this can be explained by the poor education of those responsible, who apparently confused *Hippocampus* (seahorse) with hippopotamus. The twelve-inch-long enamel eyeteeth of the hippopotamus were in great demand at the time. This costly material was used in manufacturing false teeth. Seahorses, on the other hand, do not have any teeth at all.

Meyer, along with his colleagues in Europe and Asia, discovered a peculiarity in seahorses' genetic material during their microbiological studies, which might explain why these marine creatures are toothless. They determined that an important phosphoprotein, which plays a central role in forming enamel, is missing in seahorse DNA. It's highly probable that the loss of this protein over the course of evolution is the reason seahorses don't have any teeth. "Why would they need teeth?" asks Meyer. The seahorse's jaws have grown together into a tubular snout, which is highly suitable for sucking up the tiny zooplankton that form its diet.

So-called genetic switches are often responsible for evolutionary changes in body structure, switching genes (or their effects) on and off—and a lot of genetic switches are missing in the DNA of seahorses. Probably that was advantageous for the evolution of their atypical body structure and unusual swimming style—characteristics that benefit the horses of the sea to this day. They are less visible in seagrass on account of their upright swimming technique. "And maybe their atypical posture and physical structure also confuses predators, whose prey usually look quite different," says Meyer.

Seahorses lack several typical genetic switches, particularly in those areas responsible for the skeletal structure of other fish (as well as human beings). The consequences are clear—whoever thinks seahorses are delicate and fragile is wrong in the case of almost all the species. These finned creatures have a hard surface—a kind of exoskeleton formed of interconnected bony plates covered in scaleless skin. On the other hand, they do have a vertebral column, yet neither any fish bones nor any ribs. The connections between the bony plates of their exoskeleton protrude like ridges, lumps, or spines and can help experts differentiate between the various species of seahorses. In the seahorse's tail, the bony plates are interconnected via highly flexible joints.

The body armor provided by the exoskeleton, as well as its thorny appearance with little muscle on its bony plates, all serve to make seahorses unattractive, poorly digestible prey. (Experts count eighty-two different seahorse predators, including predatory fishes, such as mackerel, tuna, or gilthead sea bream, as well as octopi and seabirds, based on the analysis of their stomach contents.)

The body armor of bony plates is also the reason zoologists formerly classified seahorses as insects or crustaceans (Crustacea). In contrast to lobsters, the seahorse's protective body armor grows steadily along with it and does not have to be shed and rebuilt from time to time, which would expend a lot of energy. Experts presume that the loss of the respective regulatory sequence for skeletal formation in seahorse DNA was beneficial in developing its body armor of bony plates.

However, perhaps the most interesting aspect for researchers, such as Meyer, is the genetic recombination of DNA, which happens repeatedly during evolution. "When such recombinations occur, one of the genes can still perform its original function, while the others can undergo mutations—spontaneous, lasting changes in the genome—which may give rise to new functions," the biology professor explains. This creates an evolutionary playground. "And that's probably how it became possible for male seahorses to get pregnant." Among other genes, seahorse researchers are focusing on the replication of the c6ast genetic sequence. Copies of this gene might help to open the seal of the brood pouch, for example, which enables the males to give birth.

Yet, why does a fish need the head of a horse? Why does it have a long snout and chubby cheeks? Why is the seahorse the only fish with a neck? Why does its head face forward and at a right angle to its body—in contrast to all other fish? And what about the crown on its head? Does this majestic headgear serve as a decoration to attract sexual partners, somewhat like a peacock's magnificent tail feathers? Probably not, because why would both male and female seahorses have the same decorative headgear? Perhaps it's just an elegantly shaped protective helmet?

The seahorse's elongated snout and chubby cheeks enable it to suck up floating prey as quick as a flash, as scientific studies have proven. The longer the tubular snout, the faster the prey the seahorse can catch. And it's probably not the unusual position of the seahorse's head that gives it an advantage over other small predatory fish (see chapter 3). Recent research also indicates that the special shape and position of its head serves as built-in camouflage, enabling the incredibly slow seahorse to catch its prey (see chapter 6). In any case, it allows the horses of the sea to communicate with each other via clicking noises—for example, to find a date to mate.

= 3 =
How It All Began

FINDING THE
PRIMEVAL SEAHORSE

―――――――

"But the sea

which no one tends

is also a garden"

WILLIAM CARLOS WILLIAMS

AN ANCIENT LEGEND of the Seri, an Indigenous tribe in Mexico, is a myth explaining the origins of the seahorse. When the world was still young, all the animals could speak, and they wore clothes like people—as did the seahorse. At that time, the seahorse still lived on land—on Tiburón Island, in the Gulf of California—and he was notorious for his bad character. Once, he provoked the other animals so much that they threw stones at him to chase him away. Out of breath, the seahorse arrived at a safe beach. He slipped out of his sandals, stuck them into his waistband, dove into the sea like a daredevil, and disappeared forever.

Legend has it that to this day all seahorses have a tiny but vital fin in the very place where their ancestor stuck his sandals into his waistband—the dorsal fin, which propels them forward in the water.

Evolutionary scientists have sought other explanations for the positions of the seahorse's fins and swimming technique. They also want to determine how, when, and where these unusual fish evolved. One of the noted experts on seahorse evolution is Peter Teske, a South African professor of marine biology from the University of Johannesburg, a dynamic man with short hair and a three-day beard. "The seahorse is unique among the bonefish," says Teske. "It is the only fish that swims in an upright position, while holding its head at a right angle to its body." Pinpointing the time when this elementary characteristic evolved is still a matter of contention. There is a missing link in this phase of the seahorse's evolutionary history: a fossil that would document an important stage in its evolution.

Due to the bony plates of their body armor and their vertebral column, seahorses would seem predestined to be found in fossil form. Astonishingly, however, very few seahorse fossils have been found worldwide—one of the many unsolved puzzles involving these mysterious creatures.

It was therefore a scientific sensation in 1978 when researchers found fossils of several long-snouted seahorses (*H. guttulatus*), a species that still exists, along the Marecchia, a river in Northern Italy. Experts, such as Teske, estimate the age of these fossils at approximately 3.1 million years. They assume, however, that seahorses have existed for much longer—especially since fossils have been found of close relatives from the Syngnathidae family (the family of pipefish, including seahorses), which are at least 15 million years old.

In 2005, around three decades after the fossils were discovered in Northern Italy, paleontologists found fossils of two much older extinct seahorse species in the Slovenian hills of Tunjice. The species they belong to are most similar to present-day pygmy seahorses (see chapter 11), yet they have longer snouts. "They are fully developed seahorses, and not a more primitive ancestor," emphasizes Teske. "This indicates that seahorses must have evolved much earlier." In addition to their exoskeleton, their tubular snout, and their prehensile tail, these fossil species also assumed an upright posture while swimming, with their heads held at a right angle. "A transitional form would not be expected to have already had an upright swimming technique," says Teske, since this precise characteristic evolved at a very late stage in evolutionary history—and it's what makes the seahorses unique in the pipefish family.

The ancient horses of the sea whose fossils were found in Slovenia probably date from around 12.5 million years ago, when today's hilly region of Tunjice was part of an intercontinental sea located between Eurasia and Africa. This was determined by scientists conducting stratigraphic investigations, a method by which the layers of sediment deposited over the course of the Earth's history are used for dating. The rule of thumb is the deeper the find, the older it must be.

The missing link that could illuminate the transition from the "normal" horizontal swimming position of pipehorses to the upright position of seahorses—which must be older than the seahorse fossils from Slovenia—has not been found to this day. "Such evidence is not absolutely necessary for understanding how seahorses evolved," says Teske. Many pipefish (Syngnathidae) have hardly undergone any changes since their earliest appearance, as documented by fossils. Therefore, it's likely that the missing link in their evolution, the predecessor of the seahorse, still has living progeny very similar to their ancestors. Teske and other researchers are focusing on the pygmy pipehorses, a group of small seahorse-like creatures, which live in algae reefs in the Indo-Pacific Ocean and the Caribbean.

Just like seahorses, pygmy pipehorses also have a prehensile tail, an exoskeleton of bony plates, and fused jawbones, which they use to suck up their prey without chewing it. Pygmy pipehorses even have a small crown on their head, and the males of this group of creatures also brood the offspring. The big difference from seahorses is that pygmy pipehorses don't swim in an upright position. "Thus, a lot would indicate that an upright swimming position developed after the pygmy pipehorses and seahorses took diverging evolutionary paths," says Teske.

Certain pygmy pipehorses, like the *Idiotropiscis* (one of around thirty generic groups of pipehorses), look very much like today's seahorses. Genetic studies by Teske's team have shown that seahorses are very closely related to the *Idiotropiscis*. The biology professor from the University of Johannesburg thus surmises that pygmy pipehorses are an important evolutionary link which led to the development of seahorses.

The *Idiotropiscis* are primarily at home in Australia. Some experts thus believe that the first seahorses evolved down under. Teske's research team has been using molecular dating, a method that determines genetic differences accumulated among living species, estimating that the last common ancestor of the pygmy pipehorse and the seahorse probably lived in the Late Oligocene around 25 million years ago. During this era, monumental tectonic changes took place in the region between northern Australia and Indonesia—the collision of Australia and New Guinea with the Eurasian continental plate—which could have had a major influence on the evolution of the seahorse as an emerging species. Particularly in the western Pacific tropical zones, the seafloor was elevated as a result of this gigantic collision, whereafter flora and fauna adapted to the newly formed expanses of shallow waters. Huge beds of seagrass spread, creating ideal habitats for seahorses. These new habitats promoted the development of an upright swimming technique, which was unique to the early horses of the sea. "Seahorses are able to maneuver very well in such environments," says the marine biologist. Among the vertical seagrass stalks and leaves, they are especially well camouflaged due to their upright posture.

Yet can that explain the evolution of a new species? Teske nods. "There is a general consensus that the most probable way for a new species to be established is when its ancestral species have become isolated," he says. Biological barriers often play a central role in this process. If the course of a river changes due to an earthquake, separating the habitat of mice, for example, into a moor and a savannah, then the individuals on the left and right riverbanks can only reproduce among themselves. At some point, those two populations will become so different by adapting to their new habitats that they no longer want to mate with each other—even if the natural barrier is eliminated, when the riverbed dries out again. But what about seahorses?

During the evolution of seahorses, no natural barrier came into being, Teske explains. The new habitat with its huge expanses of seagrass, which developed during the Oligocene, gave the upright-swimming seahorses a chance to benefit from their advantage and establish themselves permanently without a lot of competition from similar fish. The *Idiotropiscis* with their horizonal swimming position remained in their former habitat of algae reefs, for the most part. "To this day, seahorses are commonly found in seagrass beds, but *Idiotropiscis* only rarely," says Teske. "For a creature like the seahorse that wants to remain hidden—from its enemies and its prey—camouflage naturally plays a central role." The beneficial environmental conditions in the seagrass beds of Australia must have supported the seahorse's natural gift for ambushing all kinds of zooplankton, surmises the researcher. Only after this new species of fish had already profited from its novel habitat did it expand around the world to various regions of the seas, where it adapted to new environments.

Another factor was important in the evolution of seahorses, emphasizes the South African professor of marine biology. The special posture of their head gives them a great advantage when hunting with their tubular snout, because it increases the so-called striking distance, or the distance

46

between the original position of the eye during the attack and the final position of the snout when the prey is swallowed. The striking distance is at least 28 percent greater in seahorses due to the position of their head—at a right angle to their body—than in the case of pygmy pipehorses, which hunt in a horizontal position, according to scientific studies. "It's possible that the seahorse's upright posture was primarily a means of maximizing the angle between its head and its body," says Teske. Even the improved camouflage among the vertical stalks of seagrass may also have been an added benefit.

An increasing number of experts are convinced that the first seahorses evolved in the vast beds of seagrass which developed between northern Australia and Indonesia during the Oligocene—this hypothesis is supported by the fact that many different seahorse species live in this area today. According to data gathered by the Canadian taxonomic expert Sara Lourie of McGill University in Montreal, at least eleven species live there. The underwater photographer and fish taxonomist Rudie H. Kuiter from Melbourne is even convinced that at least thirty different species of seahorses are at home in Australian waters.

Yet was the cradle of the seahorses really down under? Isn't it more likely that this genus evolved in several places at the same time? How else could these slow swimmers have spread around the whole world?

Most female fish spawn, releasing hundreds of thousands, or even millions, of eggs into the water. After the eggs are fertilized by the males, they drift for weeks in all directions, depending on the ocean currents. On the other hand, seahorse babies develop in their father's brood pouch, protected from the waves and ocean currents—and male seahorses only have tiny fins. Nevertheless, some of these fish must have traveled very far, according to scientific studies. From countless molecular genetic examinations, it is known that the local population of a species that lives closely together is genetically more similar than those whose habitats lie far apart from each other. However, this basic rule cannot be applied to seahorses. Researchers have often come upon astonishing genetic similarities between the local populations of a species that live thousands of miles apart from each other. It's very likely that individuals or small groups of seahorses have traveled huge distances and colonized new territories. But how can that be?

= 4 =

Bringing Up Baby

STAGES OF LIFE, FAVORITE HANGOUTS, AND MOBILITY

———

"Maturity is a high price to pay for growing up."

TOM STOPPARD

SEAHORSES DO NOT have single offspring. Each brood normally comprises around one hundred to five hundred lively baby seahorses, yet some species may even have up to two thousand. Considering all the dangers awaiting baby seahorses in the water, it's necessary to have a lot of offspring all at once. Especially since newborn seahorses, which are often as tiny as fleas, must immediately rely on their own devices to make their way in life. Their parents do not take care of them in any way. If a baby seahorse doesn't look out for itself, when a hungry predator fish swims by or the seas are stormy, it won't live for long. Often just one in two hundred baby seahorses survives until adulthood, since completely inexperienced sea foals make perfect prey for the countless zooplankton feeders.

Seahorses are highly fertile creatures. Many succeed in reproducing more than once, and each time they may have hundreds of offspring. Yet not only do many baby seahorses get consumed by predators at an early stage, but seahorses generally don't live very long lives. The two seahorses that are at home in European waters, short-snouted (*H. hippocampus*) and long-snouted seahorses (*H. guttulatus*), have an average life expectancy of around four years. Other, larger species can live up to five years in the sea, but the smaller species only survive for around twelve months.

Seahorse babies are more slender in general than their parents, whom they closely resemble. The first heroic deed a seahorse baby must perform is making its way to the surface of the sea. Not for the purpose of breathing, as people used to think, but to fill its swim bladder with air. This organ enables it to float in water without expending energy. After seahorse babies have succeeded in tanking air, then they can begin hunting for prey. However, their first excursion is risky. If the sea foals are not extremely careful, they'll be caught in the surface tension of the water, making them incapable of submerging themselves. Sometimes their swim bladders may take on too much air, and then they bob on their backs on the surface of the water until they perish.

The impressive head, dorsal, and pectoral fins are fully developed in newborn seahorses, which are only a few millimeters long; however, their prehensile tails are relatively short and weak. During their early lives, many species drift around as plankton in the water. Experts speak of a "pelagic life stage." Yet this particularly risky phase in the life of a baby seahorse only lasts around fourteen to twenty-eight days in most species. Other reef fish remain in the pelagic stage for up to four months. "During the pelagic stage, the seahorse babies suck up tiny zooplankton with their tubular snouts, but they are also preyed upon in great numbers by larger plankton feeders," says Daniel Abed-Navandi from the House of the Sea aquarium. The biological reason for this sensitive stage is related to the evolutionary requirement for dispersion and mixture of the gene pool—and serves to prevent inbreeding. A younger seahorse has a better chance of finding a mate from farther away—and that seems to be direly needed, because the territory of an adult seahorse often comprises just a few square feet of seafloor, which it will not leave during its lifetime.

Aquarium studies have shown that the horses of the sea undergo three different phases of growth. They develop fastest during the second phase, once they've let themselves sink toward the seafloor, which happens in most seahorse species after, at the latest, four weeks. During this phase, their prehensile tail length increases quickly and, as a result, most species attach themselves to a holdfast as often as possible, only rarely swimming around.

In the third phase, the juveniles of almost all seahorse species find themselves a fixed territory on the seafloor, in highly diverse habitats. It's often said that all grown seahorses live in seagrass beds, yet that's a myth. "That false belief has come about because they are sometimes sighted by divers in seagrass," explains seahorse expert Rudie H. Kuiter from Melbourne. In deeper water, these well-camouflaged creatures are simply much harder to spot. And dragnet fishers are supposed to completely avoid the coral reefs that are home to many seahorses. Tropical seahorses more commonly live on coral reefs or among sponges than in seagrass, according to marine pet shop owner Elena Theys. "That's why those seahorse species have such brilliant colors." And these nifty creatures even know how to avail themselves of products manufactured by human beings. For example, some horses of the sea curl their tails around the protective netting that has been put up along the coasts to keep sharks away from the beaches.

Most seahorse species have their habitats in the shallow water of the tidal zone. However, during the cold season, some of them temporarily move to deeper regions of the sea. "The winter months are particularly cold on the water surface, and often stormy as well," explains Daniel Abed-Navandi. "That's why some seahorses prefer to spend the winter in greater depths." In deeper waters, the temperatures are more moderate than at the surface of lagoons or shallow bays, which the creatures prefer in the summertime, and the water currents are also weaker. Three-spot seahorses (*H. trimaculatus*) are often pulled up by bottom trawlers fishing at depths of up to 130 feet in the winter in Southeast Asia, and the paradoxical seahorse (*H. paradoxus*) from Australia seems to be an exclusively deep-sea fish.

It has become increasingly clear that certain seahorses prefer different habitats depending on their stage of life or their current activity. Some species reproduce in shallow water, although they live in deeper waters. Tiger tail seahorses (*H. comes*), for example, like to hang out around coral as adults; however, during their juvenile phase, they prefer habitats of brown algae (*Sargassum*). In many regions of the world, the habitats seem to be divided among different seahorse species. In the Philippines, tiger tail seahorses live on coral reefs; however, yellow seahorses (*H. kuda*) live in mangroves or river estuaries. In Europe, long-snouted seahorses (*H. guttulatus*) prefer areas of thick seafloor vegetation, whereas the short-snouted seahorses (*H. hippocampus*) prefer scantily vegetated areas with a sandy seafloor.

Most seahorses will grasp anything they can find as a holdfast—yet it's usually something attached to the seafloor. However, certain species are particularly fussy. They prefer sponges, even if their habitat is dominated by seagrass. Bargibant's seahorse (*H. bargibanti*) finds a holdfast on a specific species of coral named *Muricella*. Hedgehog seahorses (*H. spinosissimus*), on the other hand, often make use of a kind of underwater taxi for short distances; they attach themselves to pencil urchins, which move around the seafloor by means of their suction-cup feet.

In general, all seahorses are saltwater fish. However, a few species do venture into the brackish estuaries of rivers or inlets. Perhaps, over the course of evolution, seahorses will someday also conquer freshwater habitats, like the several species of pipefish that make their home there.

In any case, the more experienced a seahorse becomes, the lower the risk that it's consumed by a predator. Seahorses must become increasingly skilled at avoiding danger, because if they're discovered by their enemies they hardly stand a chance of fleeing. Therefore, young seahorses learn to cling on to their holdfasts so tightly with their prehensile tails that only extremely strong predators can tear them off—predator fish, such as snappers or flatheads, and certain penguins.

Predators are also one of the major reasons why seahorses don't live in schools, like herds of horses on land. Although most seahorses have a steady partner, they also spend a lot of time alone. Yet why don't they gather in a school, like many small fish do, to protect themselves from danger? "In order to ambush their prey unexpectedly and to protect themselves from predators, seahorses are very reliant on camouflage," says seahorse expert Sara Lourie. "And that works best alone." Seahorses prey on tiny crustaceans and other zooplankton drifting along via the ocean currents. In a single day, seahorses can consume several thousand of these tiny creatures. They have such an enormous appetite that baby seahorses consume countless tiny zooplankton in the first hour after they're born.

In contrast to many other fish, the horses of the sea do not exhibit typical territorial behavior. They don't attempt to defend their home turf against intruders. Many seahorses spend their entire lives in a very small area. This might comprise a few dozen square feet, in the case of White's seahorse (*H. whitei*), or just a single sea fan, in the case of Bargibant's seahorse (*H. bargibanti*). And many seahorses appear to be creatures of habit. Denise's pygmy seahorse (*H. denise*), for example, always returns in the evenings to the same holdfast on the finely branched gorgonian sea fan that it traverses during the daytime.

Nevertheless, increasing evidence points to the fact that seahorses can also travel across huge distances in the sea. The pelagic stage of the juvenile creatures is too short to explain this, according to the experts. So how did seahorses manage to disperse themselves throughout almost all the seas of the globe?

The surprising reason for their high mobility would seem to be their tenacity as creatures of habit, especially in maintaining their favorite holdfasts. If a seahorse feels threatened, it instinctively clings on to its holdfast even more tightly. This is a fatal reflex, if, for instance, the holdfast happens to be a fisher's net. However, if the holdfast is algae or a twig or piece of wood drifting in the water, then it can serve as a ferry for the seahorse. Over the course of storms and via ocean currents, seahorses can travel long distances. "This has presumably enabled seahorses to cross some of the largest oceans," says marine biologist Ralf Schneider from GEOMAR in Kiel. "Sometimes parts of plants even form 'floating islands,' along with wood or other flotsam, which are used by many fish as protection," including seahorses.

By these means, seahorses can be transported for hundreds of miles on an all-inclusive trip, since some horses of the sea feed on the same miniature shrimp that live in floating algae carpets. Researchers have calculated that a single male slender seahorse (*H. reidi*) would be able to colonize a new territory after crossing the sea with such a ferry, while carrying up to 1,600 offspring in his pouch.

Further confirmation of the ferry-crossing theory is the fact that a lined seahorse (*H. erectus*), a species at home in the western Atlantic, has recently been sighted in the Azores, almost 2,500 miles away.

Some riddles in connection with seahorse mobility remain unsolved. For example, in the winter of 2006, hundreds of seahorses were washed up on the beaches of southern Australia. No one knows what led to this mass stranding. It seems to be clear that they were neither short-headed seahorses (*H. breviceps*), nor big-belly seahorses (*H. abdominalis*), which are the two most common species in southern Australia. So perhaps the creatures came from afar? The researchers also found algae between the stranded seahorses. Some experts assume that the seahorses were washed ashore after having traveled there from far away. But why? Did they arouse the wrath of the sea gods?

= 5 =

How Much Horsepower Does Poseidon's Chariot Have?

MYTH AND POPULAR CULTURE

"I'm H_2O intolerant."

SHELDON, a young seahorse, *Finding Nemo*

THE OLDEST IMAGES of seahorses in the world were created by the Indigenous people in the north of Australia—as cave paintings in Arnhem Land. According to their mythology, the spiritual ancestors created the world in a distant "Dreamtime." In the ancient legends, dances, and songs of the Indigenous peoples—and in the paintings on cliffs and the walls of caves—seahorses have survived till this day.

A mighty spiritual ancestor, known as the Rainbow Serpent in legends, plays a central role in their creation mythologies. Yet is it really a serpent, from a biological point of view? Some archaeologists have grown skeptical in the meantime. They've noted that this spiritual ancestor is not portrayed as a crawling reptile. Its curved, segmented body, tubular snout, and head held at an angle to its chest are all more reminiscent of a seahorse. Some of the "Rainbow Serpent" cave paintings even have a protrusion on their belly and could be pregnant stallions of the sea. Pure coincidence?

During a closer analysis of the over-six-thousand-year-old cliff paintings, the experts found countless motifs that cannot have been inspired by flora and fauna on land, such as sea urchins, seagrass, and sea cucumbers. The period when the paintings were made also had a direct connection to the sea, explains British marine biologist and science writer Helen Scales in her book on seahorses, *Poseidon's Steed*. The Ice Age ended around ten thousand years ago in Australia. "Ice sheets melted and poured into the oceans," writes Scales, "pushing up sea levels and causing shorelines to race inland. Old landscapes were drowned and new ones created. Treacherous, swelling seas and stormy weather would have... thrown rainbows into the skies." It's even possible that bizarre creatures of the sea, such as seahorses, were also cast ashore, according to the marine biologist. It was right at the end of this period that the Indigenous peoples created the relevant cave paintings.

Some archaeologists also interpret the cliff paintings as a reaction to the climate change taking place at that time in history. Perhaps the artists or shamans who painted them were also trying to come to terms with their radically changing habitat—and their new impressions were portrayed in connection with their ancient spiritual ancestors, such as the Rainbow Serpent.

In Europe, on the opposite side of the globe, the oldest portrayals of seahorses were found on the Mediterranean island of Crete. They're from the Bronze Age, making them around one thousand years younger than the Australian cave paintings. The Minoans, who lived around five thousand years ago on Crete, cultivated grapes and olives and shepherded their flocks—and were also known for their talent at blacksmithing. Among other artifacts, a three-faced prism of bronze has been found decorated with two lifelike seahorses closely entwined, like the famous black-and-white yin-yang symbol of the ancient Chinese empire, with the two creatures portrayed head to toe. Scholars surmise that real seahorses inspired the Minoan blacksmiths to create this motif.

Soon after this piece of jewelry was created, more abstract seahorses also became highly popular in early European art— hybrid creatures with the head and forelegs of a horse and the body of a fish. Such a chimera was called a *hippocampus* in ancient Greek, combining the words *hippos* for "horse" and *kampos* meaning "sea monster."

In ancient times, the sea god Poseidon was himself considered the supreme master of the seahorses—a bearded giant with a trident, dwelling in a shimmering palace of coral deep under the waves of the sea with his wife, the beautiful

Amphitrite. Poseidon was just as volatile as the sea itself. He often traveled through the waves in a golden chariot to safeguard the welfare of his empire. His noble chariot was pulled by a team of two to four powerful hippocampi. Fishers in ancient Greece believed that the seahorses they found as bycatch in their nets were relatives of the fiery hippocampi from Poseidon's stud farm. And the legend had further implications. In ancient times, to make an offering to honor Poseidon, the Greeks regularly killed horses by drowning them in the sea.

In many ancient cultures around the Mediterranean, hippocampi played a role—for the Phoenician merchants and seafarers in Asia Minor, as well as the Etruscans in central Italy. Hippocampi are often depicted on the walls of their funerary chambers or on their sarcophagi, surrounded by other creatures of the sea. The Etruscans particularly, who had their cultural apex from 800 to 350 BCE, often decorated their burial chambers with seahorse motifs. But what did the horses of the sea symbolize in ancient funerary art?

Originally, historians of ancient times assumed that the hippocampi were meant to accompany the dead on their journey to the underworld. However, several portrayals dating from the Roman and Etruscan eras have given rise to another interpretation. They depict warriors fighting hippocampi. Perhaps the ancient Romans and Etruscans considered seahorses to be monstrous gatekeepers, which the souls of the dead had to overcome on their journey to the underworld.

The so-called horse monsters of the sea never entirely replaced the naturalistic seahorses portrayed in paintings, sculptures, and the applied arts. The myriad hippocampi galloping through ancient Greek mythology achieved international fame, especially after the seventh century BCE, when playwrights and poets began appropriating the oral myths, and Greek gods became known all over the Western world. Although seahorses don't play a leading role in any of the ancient myths, the horse teams pulling Poseidon's chariot, as well as the mounts of the sea deities (the Nereids), were highly popular, as documented by portrayals on numerous murals and vases.

Beginning in the third century BCE, a new empire came to power in the Mediterranean—the ancient Romans. The Greek city-states were at war against each other and were further weakened by insurgent Celtic warriors. Over the course of long years of war, the Romans gained power and expanded their empire into Egypt, Syria, and central Asia. By around 100 BCE, the Roman Empire had become the greatest world power. What about the world of Greek gods? The Romans simply adopted them. Poseidon retained his trident, his long beard, and his fiery constitution—as well as his team of hippocampi. The only novelty: the Romans called him Neptune.

The horses of the sea enjoyed an unfailing popularity, and it's quite possible that the Roman hippocampi formed the basis of the seahorses that appeared in art in other regions of Europe several centuries later. Roman troops occupied England and ventured even farther northward. In what is now Scotland, they encountered the warriors of the Picts. The art and culture of this people included animal figures hewn of stone—and some of these look a lot like seahorses.

The Picts created impressive works of art. They knew about the "golden ratio," the mathematical formula according to which the Great Pyramid of Giza and the Acropolis of Athens were built. One of the most beautiful seahorses from the era of the Picts was found in 1986 on the Orkney Islands of Scotland—an elegant portrait hewn from stone, around two hands high, with the characteristic tubular snout and curled prehensile tail.

Other charming portrayals of seahorses are found in the courtyard of the church of Aberlemno, not far from Dundee. Between Christian gravestones, there is a large stone depicting battle scenes. Ancient historians surmise that it shows the victory of the Picts over the Angles in 685 CE. On the reverse side, there are two stylized creatures with horse heads, forelegs, and fish tails. Did the Roman hippocampi serve as models? British marine biologist Helen Scales thinks these mythological creatures were probably inspired by living seahorses washed up on the Scottish coast. Whatever the case may be, she says the dozens of ancient sculptures all over Scotland makes it clear that people there have been captivated by seahorses for a long time—both real and mythological.

The belief in sea gods and seahorses outlived the Roman Empire. So-called bestiaries became popular in the Christian lands of Western Europe, as well as in the Muslim territories of North Africa and the Near East. Many people thought that each creature on land had a pendant living in the sea and, in addition to seahorses, sea goats, sea lions, and sea cows were listed as having fish tails. For example, the encyclopedia

Hortus sanitatis from the fifteenth century included several woodcuts of seahorses—as pendants to the various species of horses living on land.

Seahorses continued to inspire people's dreams—and nightmares. In Celtic folklore in the Middle Ages, the greatest sea god was named Manannán mac Lir. He also appears in various forms in Scottish and Welsh legends. Yet all these myths have something in common. Just like Poseidon, Manannán mac Lir rode a chariot pulled by seahorses. The most famous of these horses, named Enbarr of the Flowing Mane, was considered faster than the wind. The Isle of Man, between Ireland and England in the Irish Sea, was also named after this sea god. In the Middle Ages, people made offerings to Manannán mac Lir on the Isle of Man in the midsummer night—and a few of the fishers there swore they had seen his favorite horse, Enbarr of the Flowing Mane, galloping in the waves of the sea.

Other legends from the British Isles tell of water spirits that can take on the shape of horses with huge tail fins. These shapeshifters, called *kelpies,* live in the flowing waters of Scotland and Ireland. If you should try to ride one, it will pull you down into the depths and try to drown you. Similar creatures are called *Bäckahästen* in Scandinavia.

In Great Britain in particular, the irrational fear of any kind of horse of the sea continues to this day. "And today, thousands of years after the first mythical sea horses were dreamed up," writes Helen Scales, "the idea of water horse spirits may even have spawned a legend that many people still want to believe: the Loch Ness Monster."

Seahorse icons—both naturalistic depictions and artistic renditions—have been popular for centuries in cave paintings, mosaics, on gravestones, jewelry, vases, stamps, brooches, and dress cloth. *Hippocampi* decorate Venetian gondolas and coins, as well as decorative washing bowls from ancient Rome. In the Middle Ages, seahorses decorated heralds and shields; in the Baroque era and in the Renaissance, they were popular in paintings. And what about in our times?

The horses of the sea can be found on the heralds of the German towns of Hiddensee, Timmendorfer Strand, and Zinnowitz, as well as numerous coastal towns in France, Spain, and Portugal. On British postage stamps from the years 1913 to 1939 they pull Britannia—the female personification of the United Kingdom—in a chariot through the waves of the sea. And seahorses even took to the skies in the twentieth century. In 1929, the first winged hippocampus took off as the logo of the airline Air Orient (and its successor Air France) in Paris. This was a very special seahorse. Its upper body was based on the winged horse Pegasus from ancient mythology, and its lower body symbolized the dragon of Annam in commemoration of the colonial territory of French Indochina. However, malicious gossips claimed that the airline was actually inspired to use this logo due to all the seahorses that were spotted in the Gulf of Naples—after an Air Orient plane crashed in that bay.

Innumerable people around the world have valued these creatures for centuries as good luck charms, particularly when traveling. In the eighteenth century, Captain James Cook's 110-foot-long sailboat HMS *Resolution* had a seahorse as its figurehead. And ever since ancient times, these special fish have been associated with transporting passengers—usually important figures such as gods. Disney's 1989 animated film *The Little Mermaid* also remains true to this tradition. Ariel's father, Triton, the god of the sea, has his chariot pulled by seahorses, just like Poseidon did in ancient times.

Even musical seahorses have come into being in our contemporary culture. The British indie band the Seahorses formed around former Stone Roses guitarist John Squire and managed to achieve international fame in the 1990s. In 1997, their hit "Love Is the Law" was number three on the UK Singles Chart. A music magazine spread the rumor that the band's name was based on the anagram "He Hates Roses," as John Squire's payback to his former band members. However, the musician made it clear in an interview that a horse of the sea inspired him to take on the name—a seahorse sculpture in a club. He had bumped into it on his way to the men's room.

We might therefore deduce that seahorses even pose a danger to people—at least if they are human-sized sculptures made of acrylic. Some postmodern fantasy figures based on seahorses are not to be underestimated—for example, two characters from the Japanese video game series Pokémon. Seadra is a stallion of the sea with poisonous, spiky dorsal fins. As with real seahorses, the males in this creation of the entertainment industry are also responsible for bearing the offspring, and they bravely defend their nests—unlike seahorses—against rivals. Kingdra, an evolution of Seadra, is even more threatening. This monster seahorse hides in caves on the seafloor and can create tornadoes. Like real seahorses,

65

it can kill almost effortlessly. A Kingdra only needs to yawn, and all hell breaks loose.

However, it is the positive image of seahorses that prevails in mythology, popular culture, and everyday life. Fishers from Vietnam swear on a special energy drink to help them through long days at sea—tiny seahorses in whisky. In the Afro-Brazilian Candomblé religion, seahorses are used to repel the evil eye. And in Asia, some pregnant women wear a dried seahorse on a cord around their neck. When their contractions begin, the finned creature is put in hot water, and a seahorse tea is made to promote an easy delivery.

In our commercialized modern world, seahorses can be found almost everywhere. Their profile decorates soap bars, towels, tiles, the facades of restaurants and bars along the seacoast, as well as medals for special achievements. There are keychain holders in the form of seahorses, as well as chocolate pralines and jewelry. Dried seahorses are sometimes used as yo-yos or fridge magnets, whereas replicas are shaken up in snow globes. The Kingdom of Bahrain plans to make an homage to seahorses in the form of an artificial island in the Persian Gulf which is meant to look like a giant seahorse from a bird's-eye perspective.

In the 2003 animated film *Finding Nemo*, seahorses, those proud creatures that pulled the chariots of the gods, were only considered good for a laugh. A seahorse appears in a small role, constantly sneezing on account of an allergy to water! May Poseidon strike down that screenwriter.

= 6 =

Lazybones Are Good Hunters

THE STEALTH TRICK

"Never confuse movement with action."

ERNEST HEMINGWAY

SEAHORSES CAN SUCK UP their prey quicker than the human eye can perceive—almost like fast food. Copepods, for example, are among the favorite prey of seahorses and only measure from around 0.2 to 2 millimeters in size, yet they are quick as a flash. When a predator approaches them, they catapult themselves out of the danger zone using their oar-shaped legs. Copepods manage to travel up to five hundred times their own length per second. If we convert that speed to the length of the human body, it would be faster than the speed of sound. In this instance, nevertheless, it's the lightning-fast copepods that are afraid of the leisurely seahorses. Scientific studies have demonstrated that seahorses successfully capture the prey they spy nine times out of ten. They hunt far more efficiently than many other predatory fish. How can that be?

One factor that plays a role is the seahorse's wonder weapon—a tubular snout, like a suction tube. Suction, fatal for the prey, is created when the seahorse depresses its hyoid bone, then twists its head fast as lightning, elevates its neurocranium, and expands its cheeks. This creates lower pressure inside the mouth cavity, causing the water—and the prey—to be sucked into the snout.

Once the seahorse is close enough to its prey, it positions itself directly under the prey, head pointed downward. Then its tubular snout suddenly darts upward, and the horse of the sea sucks up a vortex of water along with the prey into its toothless mouth. According to the reports of olden days, the seahorse's cheeks developed such a force that pulverized feed was emitted from its gills like clouds of smoke. Perhaps the seahorse's spectacular feeding method also gave rise to the formerly common belief that these creatures were fire-breathing sea monsters.

In recent times, researchers have studied the seahorse's feeding method with special monitoring equipment and determined that its speed counts among the fastest recorded in the class of vertebrates. One puzzle remained unsolved for a long time: How could these slow swimmers get close enough to their lightning-fast prey to successfully capture it?

When a predator approaches, copepods usually notice it in advance due to its so-called bow wave. Copepods have poor vision, so they sense danger through the sensory cells in their epidermis. Many other marine creatures also have poor vision, yet they have special structures enabling them to sense tiny undulations—for example, fish have a lateral line organ, which can sense motion in all directions. No matter whether the enemy is approaching from above or below, in front or behind, it will give itself away by the undulations it creates. When copepods notice a minuscule change in the water pressure approaching them, they flee, so a seahorse wouldn't have a chance to catch them. Experiments have demonstrated that the seahorses, if they're to be successful, cannot catch their prey until it's just a few millimeters in front of their snout—yet the bow wave *should* give them away long before that.

The team of scientists working with American biologist Bradford Gemmell of the University of South Florida in Tampa has examined the tricks that seahorses use to take their prey by surprise. For their investigation, they chose the dwarf seahorse (*H. zosterae*), which swims in slow motion and holds the record for the slowest fish in the world. In experiments, the researchers put either a dwarf seahorse or another fish in a water tank where copepods were being kept. Using several cameras, they filmed the predators during their hunt and minutely analyzed the currents they created while approaching their prey. Based on the motion of tiny, suspended particles in the water, a computer program calculated a three-dimensional image of all undulations created in the water surrounding the predators.

At first, the researchers were not able to determine any difference between the seahorses and the other predators.

They measured relatively strong, conspicuous vortices near the head and torso. However, one area was unaffected. Diagonally above the snout, there was a small, mainly undisturbed zone in the water. Even when a seahorse approached a copepod relatively quickly, the hydrodynamic disturbances in this calm zone were below the level that would have triggered a flight response in the prey. Over the course of evolution, seahorses must have adapted perfectly to make practically no bow wave diagonally above their snout.

How is this mysterious calm zone achieved in the bow wave? One explanation came to light via further experiments done with artificial models of seahorse heads, as well as those of other predator fish with typical head shapes. These investigations demonstrated that the seahorse's very thin, long tubular snout is the decisive factor. Gemmell explains that it enables the water to flow past more easily, and with less turbulence than would be the case with a shorter shape. And since the seahorse's little mouth is right at the end of its very slim snout, it has already come dangerously close to the prey before the latter realizes it's being targeted.

Dwarf seahorses apparently use a stealth technique not unlike that of the engineers who constructed stealth bombers. Military jets use this method to avoid being detected by radar. To avoid reflecting electromagnetic radar waves back to the sender, the surfaces of stealth bombers are angled to create as little resistance as possible—just like the seahorse's snout in water.

However, stealth stalking only works if the angle is just right, as demonstrated by the experiments in Florida. If a seahorse approaches its prey at an angle that's just a bit too steep, the copepod will be in a zone where the water turbulence is significantly stronger, and the prey will be long gone before the seahorse can catch it. But the seahorse almost always succeeds. "Seahorses have the capability to overcome the sensory abilities of one of the most talented escape artists in the aquatic world—copepods," says Gemmell. "People often don't think of seahorses as amazing predators, but they really are." And that's a good thing, since they're always hungry. Seahorses need to feed constantly, because their rudimentary digestive system is very inefficient.

Seahorses are highly gifted hunters. They are aided not only by their stealth technique in hunting copepods. They have neither claws nor teeth, but they do have 360-degree vision. Just like chameleons, their eyes can move independently of each other, thus enabling them to look in two directions at the same time—ideal for spotting prey in any direction. If needed, one eye can spy for prey while the other is on the lookout for predators.

Seahorses have a first-class sense of vision. Even tiny creatures moving very quickly can be localized visually without the seahorse moving its body or head. The funnel-shaped fovea centralis, the area on the retina with the sharpest power of vision, also has a particularly high density of photoreceptor cells. Experts presume that this feature enables the seahorse to enlarge the images it perceives, like zooming a camera, ideal for a predator that very often ambushes its prey.

Most seahorses live according to the smart, sustainable motto: less is more! At least in relation to the energy they expend while hunting. They chill out until nature delivers them their dinner. With infinite patience, and perfectly camouflaged, most seahorses simply wait until their potential prey floats toward their snout via the sea currents. Without wasting their breath or straining to swim, they manage to swallow, quick as a flash, thousands of tiny crustaceans, such as mysids and water fleas, and fish larvae each day. Tiny fish also count among the prey of certain seahorses, as well as snails—and some species may even feed on baby seahorses. On occasion, they also feed on vegetarian fare. However, zoologists surmise that this isn't intentional. Seahorses are likely to accidentally swallow bits of aquatic plants as bycatch while hunting their live prey.

Some seahorses ambush their prey on the seafloor. Others snap up their prey from the water surface, while yet other species rely on refined tricks, such as the stealth technique of the dwarf seahorses (*H. zosterae*). Three-spot seahorses (*H. trimaculatus*) squirt a water jet into the sediment of the seafloor, stirring up the tiny invertebrates hidden there to capture them. "If they suck up any sand, it is expelled via the gills," explains Sara Lourie. Their prey, which is usually a good bit larger, is retained in their gullet.

Long-snouted seahorses (*H. guttulatus*), on the one hand, often let themselves sway back and forth in the sea currents, entwined on a stalk of seagrass, leisurely snapping up zooplankton when it passes by. On the other hand, short-snouted seahorses (*H. hippocampus*), are more active. They ambush their prey and prefer to snap it up off aquatic plants.

Even though many marine creatures do not have eagle eyes, visual camouflage plays an important role for seahorses in hunting mode. Finding seahorses in the wild is very difficult, writes marine biologist Helen Scales: "One can be right in front of your dive mask and still you won't see it as it hides beneath a shroud of crafty camouflage." Many seahorses can change color when needed, enabling them to blend in with their background. Some look as though they were leaves, stalks, or seagrass. "They can adapt very closely to various habitats by changing their color—for example, [mimicking] red algae or orange-colored sponges," says Daniel Abed-Navandi, deputy director of the House of the Sea in Vienna. "They do that via multi-colored cells [chromatophores] in their skin, which can expand, shrink, or overlap to create changes in color." For example, going from black to bright orange. Seahorses particularly loyal to their immediate habitats allow algae, bryozoans, or small freshwater polyps to attach themselves to their surface. Special cells emit a sticky substance, assisting this process that enhances their camouflage abilities.

The best camouflage artists are the tiny Bargibant's seahorses (*H. bargibanti*), whose exclusive habitat is on *Muricella* gorgonian sea fans. With their tiny, wartlike skin bumps (or tubercles) and their extremely short snout, they look just like this specific fan coral, resembling a bush with springtime buds, and not only in terms of color. Even the shape of their body imitates the structure of the coral branches. Bargibant's seahorses camouflage themselves in purple with pink or red knobs on *Muricella plectana*, and yellow with orange-colored bumps on *Muricella paraplectana*.

In summary, seahorses may not be as famous for their hunting prowess as tigers, wolves, or sharks. However, their wonder weapon, a virtuoso stealth technique—in addition to other nifty tricks—enables them to polish off thousands of tiny zooplankton daily. Both the stallions and mares of the sea hunt and feed almost constantly in the wild, unless they are dancing in preparation for their wedding night—or if the males have become pregnant again.

= 7 =

Chatting

WHY SEAHORSES CAN'T KEEP
THEIR MOUTHS SHUT

―――――

"If you were to make little fishes talk,
they would talk like whales."
OLIVER GOLDSMITH, to Samuel Johnson

JOHANN WOLFGANG VON GOETHE, who is often called Germany's greatest poet, also considered himself a gifted natural scientist. However, he seems to have known little about fish—at least about seahorses—as he once wrote, "Water alone makes one mute, as seen with the fish in the river." But these finned creatures, despite what Goethe thought, are not mute at all.

The biologist Karl-Heinz Tschiesche—head of the marine aquarium in Stralsund, in former East Germany, from 1972 to 2003—had already gleaned that fact from expert publications before the Iron Curtain fell. The scientist confirmed that seahorses don't whinny like horses on land. "Yet, during their courtship, they produce sounds that remind one of drum-rolls." At least that's what Tschiesche had read in scientific publications. Underwater drumrolls? He was a bit skeptical. Eventually, he decided to conduct a thorough investigation.

A physicist built him a hydrophone, an underwater microphone. Over many long hours, Tschiesche waited beside the marine aquarium, a loudspeaker next to his ear and the hydrophone attached to a cord swinging in the water. Yet he couldn't hear the seahorses make a single sound. "Couples were forming in the tank at the time," recalls the researcher, who has a long white beard and eyes sparkling with curiosity. "Actually, it would have been the ideal time to hear their ominous courtship drumrolls." Sometimes, it seemed to Tschiesche that the male and female seahorses really were flirting with each other. "As soon as I moved the hydrophone near to their tubular snouts," he says, "they turned around, swam away, and hid behind a stone." The creatures often had their prehensile tails intertwined, the faster seahorse pulling the slower partner along behind it.

Finally, the head of the Stralsund marine aquarium put his hand into the tank, grabbed one of the seahorses, and put its tubular snout in front of the underwater microphone. "Promptly, the fish emitted a long series of sounds," explains Tschiesche. "I'm sure it wasn't the coveted love song that it was singing at the time," he says, smiling. The calls didn't sound like drumrolls—rather more like a low buzz-ing noise. "At least the seahorse was making noises that it presumably would have made during other natural states of excitement."

Many years have passed since then, and the Iron Curtain has fallen. Tschiesche is now in his eighties, and investigating the sounds made by fish has become an independent field of research. Biology professor Friedrich Ladich at the University of Vienna has been researching this topic exclusively over the past thirty years. "It isn't just a handful of chatterboxes that quack, whistle, buzz, call, peep, caw, and trill in the seas, lakes, rivers, and aquariums of the world," says the bioacoustics expert. "Researchers now estimate that up to 50 percent of the thirty thousand known species of fish communicate via acoustics." Most of the sounds they make are between the frequencies of fifty and eight hundred hertz, which can also easily be perceived by the human ear. "In water, sound waves spread four times faster than in the air, and can travel greater distances," explains Ladich.

In ancient times, a few scholars were aware that fish were anything but mute, he says, taking a well-read book from a bookshelf—*Historia animalium* (*History of Animals*) by Aristotle, written in the fourth century BCE. "Fishes can produce no voice, for they have no lungs, nor windpipe and pharynx," he reads. "But they emit certain inarticulate sounds and squeaks, which is what is called their 'voice.'" The "sciaena," for example, which "makes a grunting kind of noise"; the boarfish; the "chalcis"; and the "cuckoo-fish." The spectrum of sounds they make ranges from a "sort of piping noise," to a call "like the cry of the cuckoo." Some fish emit noises that sound like language "by a rubbing motion of their gills," postulated Aristotle, others "by internal parts about their bellies." Modern research has not been able to confirm his theories about gills and fish bones near the stomach. However, rubbing does play a central role in many sounds created in the realm of fish. "The scientific term is 'stridulation,'" says Ladich, closing the covers of *Historia animalium.*

The stridulations emitted by some fish are created, for example, by rubbing together hard body parts. Some fish species scratch the rays of their fins together—the mainframe of their propulsion organs—while others pluck the tendons of their bodies like the strings of a musical instrument. Others grate their teeth loudly. Another large group of fish uses their swim bladders, the organ which helps them regulate their buoyancy. Many species create low-frequency sounds, making the gaseous content of their swim bladder oscillate by rhythmically contracting their so-called drum muscles quick as a flash; some species even manage 250 contractions per second. This creates low, rumbling, purring, or honking noises, such as those typical for cod, explains Ladich.

Next, the bioacoustics expert opens the door to a laboratory, where numerous aquariums can be seen. He doesn't even need an underwater microphone to hear some species chatting or quibbling. "Listen for yourself," says Ladich, fishing out, with a landing net, a brown and white striped catfish. The striped Raphael catfish, around eight inches long, tries to flee and makes a hoarse croaking noise.

"Some fish are real communicative geniuses," says Ladich, releasing the fish into the water again. Some of them relay their intentions and emotions by changing color. Other fish use scents, while some use electric signals, and many use sounds. The most important functions of the acoustic fish language are to find a partner and mark a territory, somewhat like birdsong. However, when fish create a ruckus in water, it might also be caused by territorial fights—or to warn against danger. Experts differentiate between mating calls, alarm calls, fighting cries, and a large repertoire of love songs.

But what about seahorses? At the turn of the twentieth century, naturalists realized that seahorses were not mute either. Some seahorses chirp like crickets. Others produce noises that sound a bit like people nervously snapping their fingers. Initially, scientists believed that the noises were produced when the creatures shook their lower jaw. However, that has been disproven in the meantime. Recent research has determined that some seahorses' decorative coronets can be rubbed against their skullcaps to make clicking noises. When scientists removed the corresponding bones in experiments, the seahorses clicked less often—yet they didn't completely stop making clicking noises. Therefore, they must also use another method to create these sounds.

Why do seahorses want to draw attention to themselves acoustically? And who talks more, female or male seahorses? Some male animals—on land and in water—seem to be even more communicative than their female counterparts. "In the realm of fish, the males make mating calls to attract females," says Ladich. Female finned creatures are usually relatively quiet in comparison. But seahorses are the exception, once again. Female slender seahorses (*H. reidi*), whose language has been studied by Ladich together with other scientists from Austria and Brazil, have proven themselves very eloquent. The researchers discovered that the males and females of this species emit twenty-millisecond-long "mating clicks." And these sounds cannot be differentiated in their length or frequency between the sexes. According to the scientists, the clicks do not serve the purpose of finding a mate—so then why are they made?

"Clicking together probably plays an important role in synchronizing the courtship behavior of pairs—in the preliminary stage before mating," says Ladich. In this sensitive period, anything helpful is worth its weight in gold. "Only if the male and female work together very closely will the female be able to successfully deliver her eggs to the male's pouch for fertilization." The fact that the clicking tempo increases the closer the pair comes to mating leads to the conclusion that it's connected to the seahorses' sexual reproduction. The production of as many offspring as possible is the main goal of every creature in the kingdom of animals.

Experts are still wondering why slender seahorses also make clicking noises while they're hunting for prey, noises that are almost double the volume of their loudest mating clicks. The purpose of these loud hunting calls remains a mystery, as they don't increase their hunting success rate. On the contrary, seahorses risk being heard by predator fish. Ladich has come up with a hypothesis in the meantime: "Maybe the seahorses want to indicate their good food sources to prospective mates."

Without a doubt, slender seahorses are not the only communicative seahorses. Dwarf seahorses (*H. zosterae*), lined seahorses (*H. erectus*), tiger tail seahorses (*H. comes*), and other species also communicate with each other via clicking noises. The sounds they make during courtship and mating, as well as when males are competing, can be recorded by researchers using special equipment. In aquariums, many seahorses also click at feeding time or when they are stressed—for example, when one of the creatures is transferred to a new tank.

Ladich also conducts experiments on the hearing capacity of finned creatures in his laboratory, where there are more than a dozen aquariums. The researcher puts electrodes on their heads to measure the currents in their brains, like the method used to assess the hearing of infants in pediatrics. He has been able to prove that many water-dwelling creatures can hear rather well.

They may not be visible, yet fish do have ears—little sacs filled with liquid behind their eyes, which function similarly to the inner ear of land-dwelling creatures. Sound waves create vibrations in the whole body of the fish, because they have a density close to that of water. Small ear stones (otoliths) of lime vibrate an instant later, which are felt by tiny sensory hair cells that relay the information to the brain. Around one-third of all fish species also have inner ears connected to their swim bladder via fine bones, so that the vibrations of the swim bladder are relayed to their ears. "Goldfish, for example, and the large array of ten thousand related species, have really good hearing," says Ladich. "They have several auditory ossicles [small bones in the ear]—like those of human beings—and also use the anterior wall of their swim bladder as a drumhead."

What about seahorses? Some scientists postulate that seahorses use the sounds they make for the purpose of orientation. Like bats, they send out sound waves which are reflected back to their ears from nearby objects as an echo. The seahorse's brain then calculates the image that the sound waves have reflected. Ladich is skeptical. The seahorse's sense of hearing is comparatively weak, and entirely reliant on otoliths. Many other fish have a better sense of hearing. And something as complicated as echolocation would almost certainly not work with the seahorse's rudimentary hearing organs. The horses of the sea seem to be loud and cheeky— even though they don't hear much of the noise they make themselves.

All the more surprising is that slender seahorses (*H. reidi*) even have a third method of making noise—in addition to various clicking sounds made during mating and hunting. If they are caught or taken out of the water, they make strange growling and buzzing noises, probably somewhat like the sound that Karl-Heinz Tschiesche heard at his aquarium in Stralsund. At two hundred hertz, these sounds have a lower frequency than the clicking noises, and they're almost certainly made in a different way. However, it's not at all clear how. Experts consider it unlikely that the oscillations are made by gas in the swim bladder. Ladich and his team have not found any so-called drum muscles in the slender seahorse that could create rhythmical contractions of the swim bladder.

The researchers are even more interested in finding out why seahorses would need to make those mysterious growling and buzzing noises. Biologist Tacyana Oliveira from Paraíba State University in Brazil, who has done research with Ladich at the University of Vienna, concluded that these sounds are too quiet to serve as a warning signal for other seahorses—due to their limited hearing capacity. On the contrary, she surmises that the growling sounds and vibrations made by stressed seahorses could potentially irritate their predators. "Maybe just enough so they can free themselves."

It's possible that if a seahorse feels its life is in danger, it might briefly surprise a predator fish or bird with its growling and buzzing vibrations, just long enough to free itself. The bad news for the horses of the sea is that bottom trawlers, which constitute the gravest danger for seahorses in our modern world, don't have any ears (see chapter 15).

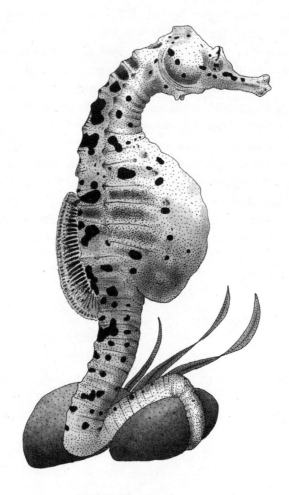

Big-Belly Seahorse
Hippocampus abdominalis

= 8 =

Underwater
Ballet

LOVE, SEX, AND PARTNERSHIP

===

"Love is an irresistible desire to be irresistibly desired."

ROBERT FROST

WITH REGARD TO couples' relationships, human beings could learn a lot from seahorses. While human partners can sometimes spend too much time together—until they feel suffocated—male and female horses of the sea aren't together all day long. They only spend quality time together. They chat, flirt, and dance together, and they mate. The rest of the time, males and females go their own way, take naps, and let themselves drift in the ocean currents—or they might hang out somewhere, leisurely eating their fill.

After they've had enough alone time, they look forward to getting together again. Seahorses are born partygoers. A seahorse stallion would never stutter to a mare, "Sorry, I can't dance," if she asks him to tango. Partner dancing is one of their favorite leisure activities. Right after sunrise, males and females approach one another, gently rubbing their noses together and then sprightly beginning to circle each other. Many of them make seductive clicking noises. The partners gracefully rock back and forth, as though to the beat of underwater music. They dance, cuddle, and schmooze together dreamily, like they've completely lost track of time. However, love can be dangerous for seahorses. During partner dancing, hormones are released that can make their camouflage fade. This causes changes in color, so their bodies begin to glow, and the contrasts in the patterns of their skin become more pronounced. Researchers hypothesize this is how seahorses signal their willingness to mate.

The partner dances also serve as a means of seduction. Before mating, courtship can take many hours. Finally, the female signals that she's ready. She swims up toward the water surface, pointing her snout toward the sky, and stretches her body out straight as a stick—a pose that is irresistible to the male. The stallion of the sea powerfully presses his chin against his chest and makes his prehensile tail open and close like a switchblade. This enables him to pump water into his brood pouch to show his beloved mare of the sea how roomy it is.

Soon afterward, the mare and stallion of the sea snuggle up together closely and let themselves drift upward. They press their bodies together so that their snouts and abdomens are touching. On account of the curves in their body posture, the space between them looks like the shape of a heart. Then, something amazing takes place. A tubular rod appears in the middle of the female seahorse's belly, which looks a little like a penis, the so-called ovipositor. At the climax of the love scene, both partners lift their heads as though in ecstasy, curving their backs, and the female seahorse transfers her eggs into the male's brood pouch, while her partner fertilizes them with his sperm.

Shortly afterward, the loving couple separates. The colors and patterns of their bodies fade again. The male shakes himself to ensure that the fertilized eggs slide into a favorable position in his pouch. His partner usually swims away to go hunting and feeding. However, for the expectant father, the difficult time of pregnancy now begins.

Scientists have long been trying to determine why it is the males that become pregnant in the realm of seahorses. Only recently have the first explanations come to light (see chapter 9). However, that isn't the only mystery in the reproductive cycle of these fish. Biologists long assumed that the males—since they take on the typical female role— fertilized the eggs directly in their brood pouch. Yet the researchers have discovered another curiosity, at least in the case of the yellow, common, or spotted seahorse (*H. kuda*). The males cannot deposit their sperm in their brood pouch, because their sperm tube ends outside of it. The long-snouted seahorse (*H. guttulatus*) and short-snouted seahorse (*H. hippocampus*) cannot fertilize the eggs in their pouch either; it would be anatomically impossible. So how does the sperm reach the eggs?

British professor of zoology William Holt of the University of Sheffield is one of the foremost experts in the world on seahorse reproduction. "There's a lot that we don't know about how fertilization occurs in these creatures," he says. "But there are some things we do know by now." In the case of yellow seahorses, the male's semen is emitted while mating, but the opening of the brood pouch is around four millimeters from the end of the sperm duct. "Hard to imagine how the sperm could swim that far without getting lost in the sea, simply because they're much too slow," says Holt. "In the meantime, we presume that while the female is transferring her eggs to the male's pouch with her ovipositor, she also collects the sperm from the seawater, so it enters the pouch along with the eggs." Holt considers it very likely that fertilization in other seahorse species takes place in the same complicated manner.

"Collecting" the sperm in the water somehow? Very bizarre. Why would seahorses spend so much energy on courtship and mating? Why couldn't female seahorses simply spawn in the water, like practically all other female fish, and let the males fertilize their eggs in the water by themselves? The reproductive behavior of the yellow seahorse seems illogical and absurd—as though a hunter were to load his rifle, only to throw it at a stag instead of shooting it.

Axel Meyer, professor of evolutionary biology at the University of Konstanz, smiles. Not all processes in nature are perfectly efficient, he says. "Evolution doesn't begin with a blank page; instead it makes do with whatever's already there." Evolution is not like "an engineer—rather, it proceeds via trial and error." Organisms have to function in each and every generation, meaning that the evolutionary process is founded on serendipitous coincidences. "That's why there are random mutations in the genome, which are passed on as preconditions." Some things can only change very slowly, and others not at all. Thus, there are still quite a few "misconstructions" in nature, says Meyer. For example, albatrosses have a wingspan of up to eleven feet, meaning they are so heavy that when they land, they sometimes break their neck. "Even the design of human beings isn't fail-safe," says the evolutionary biologist. "The close connection between our airway and our esophagus leads to the risk of choking to death, for example."

It's not contradictory that Charles Darwin, founder of the theory of evolution, wrote about the "survival of the fittest" in 1869, a term that he adopted from British philosopher and sociologist Herbert Spencer. "Darwin applies this principle to specimens of the same species," explains Meyer. "And you shouldn't let the word 'fittest' make you think of a fitness studio." In nature, the deciding factor is how many offspring a specimen has, in comparison to the reproductive success of its contemporaries. "Fitness is measured as the relative chance of survival and reproductive success of the specimen's genome in comparison to its competition in the given population."

Evolutionary scientists, such as Meyer, consider the fact that the opening of a male seahorse's semen duct is not well positioned as proof that male pregnancy evolved in small steps—and wasn't planned on the drafting table.

An even greater mystery is posed by the seahorses' monogamy. Only 3 percent of all mammals have lasting part-nerships, and even fewer amphibians, reptiles, and fish do.

However, most seahorses are model couples that are never unfaithful to each other. Not only do they have a loving courtship, but they also invest a great deal in their relationship, spending a lot of time together in common activities. In many seahorse species, the couples get together daily to greet each other and dance. Behavioral scientists assume that the horses of the sea consolidate their relationship with these rituals, whereby sex plays a lesser role. The phases of the moon also seem to influence seahorse love life. During the new moon and full moon, these creatures get together to flirt more often than usual. Pairs usually remain together for life. Old tales relate that if one of the partners is caught in a net, the other won't leave it behind—instead voluntarily following it into captivity. Aquarium keepers recount that after the death of a seahorse, it isn't rare for the remaining partner to also perish within the next few days.

However, since it's the male seahorses that become pregnant, why don't the females distribute their eggs among as many partners as possible, like most males in the kingdom of animals do with their sperm? Some primates, dogs, and whales even have a kind of bone in their penis that creates a permanent erection, meaning they can copulate with several different partners one after another. In complete contrast, most female seahorses remain loyal to their partner and do not mate again until their partner has given birth and is ready to become pregnant once more. Why are most seahorses loyal till death?

"In general, we don't know a lot about the genesis of monogamy in the kingdom of animals," says evolutionary biologist Anna Lindholm from the University of Zurich. However, it is well documented that the phenomenon independently came into being in several species over the course of evolution. A recent study conducted by scientists in the United States has shown that the activity of twenty-four genes in the genome of prairie voles, poison dart frogs, and water pipits is strongly linked to monogamous behavior. Presumably, this pattern may also be found in the seahorse genome. Yet this discovery still cannot fully explain the purpose of monogamy.

"One reason that lasting partnerships can pay off is that, in some species, the offspring will only survive to adulthood if both parents care for them," explains Lindholm. For example, in storks, wolves, and beavers. That explanation may be clear, but is not valid for seahorses, since their babies swim off on their own after they're born and aren't supported by their mother or father in any way. Are seahorses simply more capable of love than other creatures?

Evolutionary biologists doubt that. Yet seahorses do have one decisive advantage due to their monogamy. Pregnant males can be sure they are carrying only their own biological offspring in their pouch, and not any so-called cuckoo's eggs that were introduced by the competition. In the case of salmon, for example, a second male often flashes by and spawns in the female's "nest," after the territorial male, or "official father," has done so, explains Lindholm.

Yet how great is the certainty of a biological pregnancy in male seahorses, if their sperm—at least in the case of yellow seahorses—isn't deposited directly into their brood pouch? Could other male seahorses also add their semen? William Holt dismisses this. Yellow seahorses' complicated reproductive strategy does give them a degree of certainty. "When seahorses mate, they're very close together." No competitor can come between them. "In addition, the male's brood pouch closes again after a few seconds, so that no more eggs can be transferred," emphasizes Holt. "It is neither possible for more than one male to fertilize a female's eggs, nor for more than one female to transfer eggs to his pouch." It's a fair deal.

The main reason many seahorses are monogamous almost certainly lies elsewhere, according to the experts. Most seahorse species live in large, sparsely populated habitats, yet they are slow swimmers and, depending on the species, some of them are very rare.

"Since it's out of the question for seahorses to go on long expeditions looking for a mate, any available sexual partner is a valuable commodity, for both the male and the female, and wouldn't be forgone on a whim," says Axel Meyer. "Thanks to lasting partnerships, fertile creatures can efficiently pass on their genes despite the low density of members of the same species." Meyer remarks on the extreme case of the anglerfish in the depths of the seas. In the darkness, it is very rare for creatures to come upon other members of their species. If a male deep-sea anglerfish meets a female, he's not fussy. He immediately attaches himself to the female, in the real sense of the meaning. At first, the skins of the couple grow together, and later even their circulatory systems. The male's jawbone regresses—and eventually it merges with the female organism. In some species of deep-sea anglerfish, the male's body completely disintegrates, except for the testes, explains Meyer. It doesn't sound like a dream wedding, but it's hard to imagine a closer and more intimate relationship.

Seahorses are usually loyal partners in love—most likely because they rarely meet other members of their own species, according to the evolutionary biologists. "Married couples" practice their partner dancing routines daily—as well as their longer wedding dances—ensuring the technically precise delivery of the female's eggs into the male's pouch. Experts surmise this also enables them to adjust their reproductive cycles, which are driven by sexual hormones. The female's next batch of eggs therefore ripens just after her "husband" has finished giving birth to the last brood.

However, as the saying goes, all roads lead to Rome—or to successful reproduction. Recent research has shown that not all seahorses are, in fact, loyal. A restless female Denise's pygmy seahorse (*H. denise*) proved that she was always on the go. Scientists observed her reproductive behavior. This tiny seahorse was two-timing her partners. She danced wedding dances with both of them and mated with both. The biological advantage was that she halved the risk of losing a mate and all her offspring. This calculating seahorse followed the strategy favored by stock market speculators and didn't put all her eggs in one basket!

Some stallions of the sea are even more promiscuous than that. Big-belly seahorses (*H. abdominalis*), which are at home in Australia and New Zealand, do not seek lasting partnerships. Instead, they mate with almost any female at hand. That fits in well with the "loyalty due to scarcity" theory of the biologists, since big-belly seahorses live together in a smaller area than most other species. Another factor also substantiates this theory. Scientific studies have indicated that many other species of seahorses—romantic wedding dances or not—have a bit on the side in the turbulence of an aquarium. If there were sufficient partners available, some male seahorses would mate with up to twenty-five different females a day. Perhaps the bon mot of *Homo sapiens*, "morality means lack of opportunity," also applies to seahorses.

William Is in Labor Again

THE MYSTERY OF MALE PREGNANCY

―――――――

"If men gave birth, they'd be less inconsiderate."
GABRIEL GARCÍA MÁRQUEZ

IT'S OFTEN SAID that men only want one thing. In the animal kingdom, there are, in fact, lots of male creatures that sow their seeds among as many females as possible, only to depart quickly after mating. However, the God of creation entrusted the males of certain species with tremendous "fatherly responsibilities." Male ostriches, for example, not only raise their own offspring, but they also raise those of other pairs in a sort of kindergarten. Some species of male frogs living in the Australian rainforest carry their babies in a pouch. And a few species of male pipefish—closely related to seahorses—schlep their growing offspring on their own bodies through the underwater world. Peter Teske from the University of Johannesburg considers these long, thin fish to be an intermediate stage in the evolution of male pregnancy in seahorses. The fertilized eggs are glued to the belly of the male banded pipefish, for example, and some species of sea dragons carry their eggs on a brood patch on the underside of their tail—whereas the greater pipefish brood their eggs in skin flaps, creating a kind of enclosed pouch. However, male seahorses have a brood pouch that's closed after conception and also takes on the function of a womb, including the placenta. The brood pouch only opens again shortly before birth.

"Male seahorses actually have a classic pregnancy," says Tony Wilson, professor of evolutionary biology at Brooklyn College in New York. In 2001, he and his colleagues were able to prove that expectant male seahorses provide nutrition for the embryos in their pouch. In the meantime, it has become even more evident how closely the male seahorse pregnancy resembles that of female mammals. For example, the male seahorse's immune system protects the embryos from infections. Scientists at the GEOMAR Helmholtz Centre for Ocean Research Kiel are currently studying the microbiome of the male's pouch (the symbiotic bacteria) involved in seahorse pregnancy. The expectant father transfers his microbiome to the embryos via his pouch, fortifying his offspring's immune systems. Furthermore, the embryos' waste materials are discharged by the pregnant male, and nutritious, high-energy fats are provided for them. A gaseous exchange enables the embryos to breathe.

Yet how does a male pregnancy take place? Once the fertilized eggs have been delivered to the male's brood pouch, the inner membrane of the pouch changes. It swells and additional blood vessels are formed. This enables the eggs to become embedded in it, in a similar way to the wall of a womb. Seahorse eggs are pear-shaped. Researchers have calculated that this unusual form results in a surface area 9 percent greater than that of a bird egg of the same volume. The larger surface area presumably means they can be more efficiently supplied with oxygen.

After around two weeks, the fragile eggshells break, and the embryos become directly embedded in the spongy wall of the pouch. The placenta of the pouch provides them with oxygen and important nutritional substances.

Ralf Schneider, expert on marine ecology and seahorses at GEOMAR, is among the few scientists who have studied these processes in detail. "During pregnancy, the circulation of the so-called pseudo-placenta in the innermost tissue layer of the pouch of seahorses and other syngnathids is excellent," he explains. "And because it is a highly permeable tissue, diffusion enables oxygen to be emitted and carbon dioxide to be absorbed in the pouch—a gaseous exchange like that taking place via the gills, but in the opposite direction."

The fertilized eggs have a yolk sac for their basic nutrition. However, the father's body also provides them with additional nutrition. The embryos receive calcium—required to build their skeleton—and lipids. Fatty lipids—such as the well-known omega-3 fatty acids—are especially important for growth and certain bodily functions. These nutrients are presumably released by glands in the male's pouch and absorbed by the embryos. "The pregnant males provide their offspring with an energy cocktail," says Schneider. And the waste materials? Soluble substances, such as ammonium compounds, are absorbed by the father's blood via diffusion and are discharged via the kidneys. Insoluble waste material does not accumulate during embryonic development. In the late stages of the pregnancy, a few days before giving birth, the expectant father will sometimes open his brood pouch a little. That is presumably to enable the baby seahorses to transition more easily to the higher salt content of seawater. "The rinsing of waste products out of the pouch is presumably also a beneficial side effect."

The results of a study recently published by the University of Sydney indicate that 10 percent of the genes that are active during the pregnancy of female mammals also play an important role in the pregnancy of male seahorses. Among others, pregnant male seahorses generate the hormone prolactin, which promotes lactation in pregnant female mammals. In the case of seahorses, the hormone regulates the nutrition provided by the male to the embryos in his pouch.

One of the most unusual characteristics of male seahorse pregnancy is the amniotic-like fluid in the pouch, where the eggs are embedded. Initially, its chemical consistency is similar to the bodily fluids of other adult seahorses. Over the course of pregnancy, however, this fluid becomes more and more like the (highly saline) salt water of their future habitat, reducing the birth trauma for the growing infants.

Yet if male seahorses can become pregnant and give birth, what makes them *male*? "Quite simple," says Axel Meyer, professor of biology at the University of Konstanz, with a smile, "their sperm." In nature, male organisms are the classic producers of a huge number of tiny, mobile sperm, whereas the larger eggs ripen in the female's body. Eggs are less mobile and take a lot more energy to be produced—also true in the case of seahorses. So far, so good.

But why don't the male seahorses swim away in a flash, instead of taking upon themselves the trials and tribulations of pregnancy and giving birth? Why don't they simply father as many offspring as possible and leave the rest of the stress to the ladies, like evolution deigned for almost all other species?

In the realm of fish, the males of many species tend to take care of their offspring, says Meyer. Some defend the fertilized eggs, and others even carry them around in their mouth, where they brood them. "With fish, it's usually the males that take care of the brood," says the biologist. For good reason. "The great majority of female fish first spawn myriad eggs in the water, and only then do the males fertilize them. This gives the females the opportunity to split the scene. They're often gone by the time the males fertilize the eggs, so that the males are left to take care of the young."

Evolutionary biologists surmise that the ancestors of the seahorses were somewhat like modern sticklebacks (Gasterosteidae), small silver-gray fish that live in schools, found throughout Europe and North America. The males of this species also take care of the eggs. In summer, when the days grow longer and the water warms up, they look for shallow water with plenty of vegetation, and their dull body color becomes bright red. The competitors wage turf wars to delineate their territories. After a male has dug out a small hollow in the loose sand of the seafloor and covered it with a woven roof of plant stems, he keeps a lookout for females that are ready to spawn. The male courts his chosen female with a zigzag dance. Using his nose, he then indicates the entrance of the nest to her. As soon as the female fish has spawned, the male immediately fertilizes her eggs. Then the female moves on, and life takes a serious turn for the male. He spends weeks protecting the nest from robbers, fanning fresh water into it and removing rotten eggs. Once the fry have left the nest for the first time, their father keeps them together in a school and makes sure they return home all together. For this purpose, he sometimes even sucks a juvenile fish into his mouth to return it to the nest.

The early ancestors of today's seahorses probably demonstrated similar behavioral patterns, according to the experts. Over the course of millions of years, some male sticklebacks may have started to carry around the fertilized eggs on their body, perhaps because there were too many predators in their territory, or not enough suitable nesting material was available. Evolutionary ecologist Olivia Roth at GEOMAR believes there may have been another reason. "Seahorses are very active hunters that regularly need to feed on live prey," she says. "That's why I think carrying their eggs around with them was an evolutionary adaptation, allowing them to feed more frequently." Slowly, as an additional protection for the brood, a pouch developed on the male. "It hasn't yet been fully determined how male pregnancy with placenta-like structures developed," summarizes Roth. "However, we presume that the loss of certain genes, which were important for the immune system, played a role."

One aspect is particularly astonishing. Why doesn't the male's immune system attack the embryos during the pregnancy, since their DNA incorporates the female's genome in addition to his own? The male organism should be able to recognize the growing embryos as foreign and attempt to fight them with antibodies, like it would fight any germ causing an infection.

"In general, pregnancies are complicated from a biological point of view," says Roth. "Basically, in mammals, including human beings, embryos should actually be rejected by the female's immune system." Why doesn't this mechanism function? That was the basic question, which Roth and her colleagues at GEOMAR sought to investigate. They thought they might be able to solve the mystery based on the special case of male pregnancy in seahorses, since their immune system needs to have undergone similar changes so that the male can take on a "motherly" role.

"We have found that over the evolutionary course of male pregnancy, those parts of the immune system responsible for differentiating between the creature's own body and foreign genetic material have changed a lot," sums up the evolutionary biologist. It appears that a creature can only become pregnant, and carry an embryo, if its own immune system has lost its sense, to a certain degree, of "mine and yours." According to other studies, female mammals maintain their own bodily immune system during their pregnancy. However, the immune system's activity is temporarily deactivated during the sensitive period. Seahorses have chosen a more radical approach to avoid rejecting their offspring, as Roth and her colleagues were able to prove—part of the seahorse's immune system has been completely knocked out. Genes responsible for the production of the major histocompatibility complex (MHC) class II molecules have undergone such major changes during evolution that they no longer function—and it is precisely these proteins that normally enable the body to differentiate between one's "own" genome and "foreign" genetic material. In human beings, MHC class II proteins are active in rejecting organ transplants, for example. "Probably it was the loss of MHC class II that made it possible for the male seahorse to accommodate the eggs in its body and provide for them there," says Roth.

This discovery was particularly surprising, since MHC class II proteins had long been considered essential in the evolution of vertebrates—as they enable a highly adaptive immune system that's suited to the present environment. "Without these genes and their corresponding functions, higher life-forms were considered a sheer impossibility," says Roth. It was thought that without the protection they provided, complex organisms would quickly succumb to infections. Now researchers are puzzling how seahorses and other syngnathids can remain sufficiently immune to infection without MHC class II proteins. It could be that another part of the immune system works more efficiently instead in these fish.

It's especially fascinating that the genes lost from the seahorse's immune system are the same genes that are attacked by the human immunodeficiency virus (HIV) in AIDS patients. The horses of the sea, which are capable of surviving despite this novel twist, could become an important model for research on immune system deficits and possible therapies—perhaps even someday helping develop new medications for HIV/AIDS patients. It's not unlikely that seahorses will become instrumental in the further development of Western medicine, albeit it in great contrast to the way they are exploited by around 1.5 billion users of traditional Chinese medicine (see chapter 13).

How do seahorses themselves profit from exchanging the typical roles of the sexes? What's the biological advantage of male pregnancy? Some scientists surmise that it enables shorter periods between pregnancies, meaning more offspring will be born in a given period. The production of seahorse eggs requires greater physical exertion by the female. However, since the male carries the embryos, the female has more time and energy to find the best hunting grounds and feed extensively, so that she can produce high-quality eggs.

As a rule, a seahorse pregnancy takes two to four weeks. For expectant males, this phase is very strenuous. In dwarf seahorses (*H. zosterae*), for example, the rate of metabolism in the male increases up to 52 percent during pregnancy. In *Homo sapiens*, by comparison, this rate only increases by approximately 20 percent, even in women in their third trimester.

For a long period of time, expectant fathers in the realm of seahorses barely move from their spot. Despite their lethargic and strenuous circumstances, the male and female seahorses still meet each morning for a short dance together. By the time the big moment arrives, the expectant father's belly has expanded like a balloon.

The female is already producing her next batch of eggs during her partner's pregnancy. This takes almost exactly as long as the length of the pregnancy, enabling the males to become pregnant again right after they mate. In particular, seahorses inhabiting warm seas, such as the big-belly seahorse (*H. abdominalis*) (which grows up to fourteen inches long), are fertile over the entire year. No one needs to ask female seahorses the touchy question that female human beings so often have to ask themselves: "Kids or career?" By nature, male seahorses are born to give birth.

More than a thousand miniature seahorses are waiting to wriggle out of the pouch of some species. They swim around restlessly in the amniotic fluid, tickling the inside of the pregnant male's belly. Then, usually at night, the contractions begin—induced, just as in human beings, by the hormone oxytocin. The male seahorse's belly cramps up convulsively. With a motion like a jackknife, he repeatedly flicks his outstretched tail against his belly to help push out the young. The wide-ranging tail movements during birth are like those made during courtship before mating, a few weeks earlier. Giving birth can take anywhere from a few minutes up to, in exceptional cases, three days. The reasons for the great differences are unknown.

Between phases of physical strain, the male seahorse pauses and seems to catch his breath, panting with pulsating gills. Researchers surmise that giving birth is painful for the male. Epidurals and caesarean sections are reserved for humans; however, seahorses sometimes also use tools to help themselves. In 1867, the theologian Samuel Lockwood described in the scientific journal *American Naturalist* how a pregnant seahorse used the help of a winkle shell. This "afforded real help in the labor of extruding the young," Lockwood writes. "With its abdomen turned towards the shell, its tail attached to the under part of it, the body erected to its full height, the animal, by a contractile exertion of the proper muscles, would draw itself downwards, and against the shell, thus rubbing the pouch upward, and in this simple, yet effective way, expelled the fry at the opening on top of the sack."

Eventually, multiple expulsions often produce clouds of delicate, transparent infants from the male's pouch—huge herds of miniature seahorses. The average size of the brood varies from one hundred to five hundred babies. However, there is a great difference among the various species. Pacific seahorses (*H. ingens*) give birth to up to two thousand baby sea foals. In contrast, dwarf seahorses (*H. zosterae*) only give birth to around ten.

In comparison to their tiny bodies, baby seahorses have large heads, apart from which they look like miniature replicas of their parents. But there's no rule without an exception: a few seahorse species have offspring which look so different from their parents that even marine biologists cannot easily recognize if they belong to the same species. Furthermore, in contrast to their parents, newborn big-belly seahorses (*H. abdominalis*), also swim in a horizontal position, just like pipefish.

What about the proud father? He doesn't get any paternity leave, neither at sea, nor in the aquarium. Not long after he has given birth, his faithful partner swims back to him, ready to mate once more. Some male seahorses already become pregnant again after a few hours. In the world of seahorses, being a father is a stressful job.

= 10 =

What Makes a Man a Man?

SEAHORSES AND EMANCIPATION

———

*"The tyranny of masculinity and
the tyranny of patriarchy, I think, has been much
more deadly to men than it has to women. It hasn't
killed our hearts. It's killed men's hearts."*

V, FORMERLY EVE ENSLER

JEAN PAINLEVÉ FROM PARIS was not only a pioneer of underwater filmmaking, but also a spearhead of feminism. He shot his most successful film in 1934—about seahorses. According to him, in this film, *L'hippocampe ou "cheval marin,"* "the balance between feminine and masculine" played a central role. In the mid-1930s, Painlevé was already calling for a new image of masculinity, before Simone de Beauvoir gave feminism a household name. "This symbol of reliability is dedicated," he said, pointing to the seahorse as a good example, "to all those seeking someone without the usual egoistical stance, as it unites masculine effort with the feminine task of caring for the brood."

The main characters of his seahorse movie were males; they were pregnant stallions of the sea. Before filming, the seahorses were brought from the seacoast to Paris in rusty buckets. Painlevé had planned to make the scene of birth the climax of his film, but he repeatedly missed shooting it. Not until he built himself a strange contraption that woke him up with a slight electrical shock if he dozed off was he finally able to film a seahorse giving birth.

Painlevé later also filmed on location at the Atlantic coast. He made a short, fourteen-minute-long, black-and-white film about the everyday life of the seahorses, including their birth. "It was lovely; the beauty of the underwater world was seductive," summarized the documentary filmmaker when he finished shooting. Yet it was also easy "to lose oneself in the depth of the water." In fact, Painlevé once almost drowned in the sea during filmmaking. The line leading from his air tank had become entangled.

L'hippocampe ou "cheval marin" is particularly interesting from a feminist point of view. As the Berlin-based science journalist Cord Riechelmann recently wrote in an homage to this film, "[it] cannot be overlooked that the courtship of the seahorses takes place without either of the sexes dominating the other." For example, it makes no difference whether the male or the female begins the daily "carousel swim, during which both of them repeatedly circle around each other, rising so close to the water surface that their crowns peek out of the water." And finally, the climax—during which the female appears to penetrate the male. With these creatures, emancipation would seem to begin long before pregnancy, according to Riechelmann.

Painlevé depicted the seahorses' courtship and mating scenes in such detail that his nature documentary film inconceivably came under suspicion of pornography, and there were calls made to censor it. When it was finally approved for screening, it immediately caused such a controversy regarding the traditional roles of the sexes in human beings that it became an icon of female emancipation in cultural history.

In contrast to some of Painlevé's other projects, *L'hippocampe ou "cheval marin"* was able to cover its production costs. In the wake of his success, the lover of the sea launched a collection of seahorse jewelry—bracelets, brooches, and earrings—designed by his partner in life and work, Geneviève Hamon, which were marketed under the label "JHP"—an abbreviation for "Jean Hippocampe Painlevé." Yet, despite their enthusiasm and creativity, Painlevé and Hamon were not good businesspeople. They lost control of their newly founded jewelry company much too quickly, and other people reaped the big profit. One might conclude that seahorses may well inspire emancipation and art, but this does not always result in long-term happiness for their fans.

Nevertheless, feminists still claim the seahorse as one of their favorite symbols. The male pregnancy of these creatures, which Painlevé already found highly fascinating, still serves as a queer model for questioning the traditional roles of the sexes in *Homo sapiens*, beginning with the basics.

"Perhaps we should think of gender as something that is imposed at birth, through sex assignment and all the cultural assumptions that usually go along with that. Yet gender is also what is made along the way—we can take over the power of assignment, make it into self-assignment," says the American philosopher and gender theorist Judith Butler.

Perhaps she was also thinking of seahorses? The females transfer their eggs while mating via a tubelike ovipositor to the male's pouch, whereas the males give birth to the babies after several weeks of pregnancy followed by contractions. That should suffice to disprove the humbug traditional biological theories employed for decades to substantiate opinions about traditional gender roles in human society— with the males working outside the home while the females care for the offspring.

Feminist ethnologist Susanne Schmitt, former director of the research association ForGenderCare, writes:

[Seahorses] are famously at odds with contemporary Anglo-American conceptions of pregnancy: males are in charge of it. Females deposit their eggs in the male's pouch, who then carries the resulting two hundred or so small seahorses until birth, when he releases them through contractions into the open waters where they disperse to the plankton layers of the oceans. The biological characteristics of pregnancy in male seahorses resemble that of female mammals, providing an example of convergent evolution, where unrelated species find similar solutions to survival's challenges. The male's pouch, just like a kangaroo pouch, provides a protective and nutrient-rich environment in which calcium, lipids, oxygen, and the right salt balance are all provided to ensure normal embryonic development.

Schmitt approves of the rhetorical connection to current discourses and concepts of seahorses—for example, the discourse regarding modern fatherhood.

This is a discourse that some people may find amusing. "The trend is for men to become motherly," scoffed science journalist Michael Miersch twenty years ago in an essay in the magazine *Mare*. "New fathers are among the most beloved media heroes. Advertisers prefer to show loving fathers splashing around with their little kids in a tub, hunting in the great outdoors, and bringing home a big check—although this lifestyle is already passé for many." At the turn of the millennium, Miersch had already diagnosed a raging "seahorse syndrome" in male *Homo sapiens*.

At least in the urban regions of the Western world, you can definitely see more fathers out and about with their children than during my own childhood. Nowadays, fathers are more likely to spend quality time with their babies and children than in the 1950s, or even the 1980s. There are worse things fathers could do. Even if a homemaker hubby may not sound sexy to everyone, male *Homo sapiens* really could learn a lot from seahorses.

Some men have even made seahorses their role model. In 2019, a debate in the United Kingdom created an uproar in the media. Multimedia journalist Freddy McConnell, then thirty-two, from the English county of Kent, wanted to have himself listed as the father on his son's birth certificate. However, this was prohibited by a court ruling, because McConnell had himself given birth to his son.

McConnell is one of the relatively few trans men who have given birth to a child. Some of them call themselves "seahorses" with a wink of an eye—including McConnell, who documented his pregnancy in a film called *Seahorse*. There are no exact statistics available on the number of trans men who have become pregnant worldwide. At present, the number is fairly low. Many doctors falsely assume that taking testosterone will render a trans man permanently infertile, but research indicates that up to 30 percent of transgender men in the United States have experienced an unplanned pregnancy. As medical research and public awareness increase, trans pregnancy is likely to occur more frequently in years to come.

Yet despite our passion for actual seahorses, how exemplary and progressive is everyday life among them? It's clear that the males do a great job of giving birth, in addition to being good dancers and loving partners. Apart from that, it's doubtful whether they would be a good role model for the new, more sensitive form of masculinity that gender theorists, among others, are yearning for. Not all male seahorses are softies. Despite seahorses' interchanged roles in reproduction, it is the males that knock heads with each other to win over the females. Conversely, the females genteelly wait at the sidelines, as though they were batting their (nonexistent) eyelashes. In fact, male Australian short-headed seahorses

(*H. breviceps*) hold regular boxing matches to impress the females.

When Jean Painlevé was shooting his cult film *L'hippo-campe ou "cheval marin,"* almost nothing was known about this. The Parisian seahorse lover, artist, and feminist, who passed away in 1989, might roll over in his grave if faced with all the new findings on the behavior of these horses of the sea.

Bargibant´s Seahorse

Hippocampus bargibanti

11

Taxonomic Jungle

HOW MANY SEAHORSE SPECIES ARE THERE?

═══════

*"Every time you dive, you hope you'll
see something new—some new species. Sometimes
the ocean gives you a gift, sometimes it doesn't."*

JAMES CAMERON

AT TIMES, BEING NAIVE can be a blessing. When the British physician John White proudly published the image of a seahorse in a journal in 1790, he thought all things about the creature could be taken for granted. For "this animal, like the Flying-fish, being commonly known, a description is not necessary," he noted below the sketch, adding that it was the familiar species of *Syngnathus hippocampus*.

If White had only known how complex the taxonomy of seahorses would become, he probably would have taken more care in publishing his drawing. However, his naiveness led to the "newly discovered" species of seahorse being named after him—the species is still known as White's seahorse (*H. whitei*).

In the meantime, the taxonomy of seahorses has become tricky even for experts. Some of them assume there are upwards of forty species of seahorses, while others claim that the genus comprises more than twice that many species. Experts have only been able to agree that seahorses are genuine bonefish (Teleostei)—akin to tuna, bass, and herring—and that they are members of the family of syngnathids (Syngnathidae, from the Greek words *syn*, meaning "with" or "together," and *gnathos* for "jaw"). There are around 230 species in this family of fish with long snouts. Most syngnathids are so-called pipefishes. These creatures swim in a horizontal position and look like elongated seahorses. Even the rarely sighted pygmy pipehorses (see chapter 3) belong to this family. They are like tiny, thin seahorses that have learned to swim in a horizontal position. The most unusual members of this family are the sea dragons. They look like large seahorses that have donned shrill carnival costumes, with wild fringes decorated with green bands or yellow dots and blue stripes. There are only three species of sea dragons, and they live off the coast of southern Australia.

Classifying seahorse species is a difficult task. These unusual fish are very different from each other in detail. However, they can still be easily confused—a strange contradiction, typical for seahorses.

No one less than the famous Swedish naturalist Carl Linnaeus laid the foundations for the classification of species in modern biology, and he was first to give an official scientific name to a seahorse in 1758. In addition to naming countless other species in the plant and animal kingdoms, he included the seahorse species *Hippocampus hippocampus*. The species is still known by that Latin name, although it's known in English as the short-snouted seahorse. Before Linnaeus's pioneering work, horrible tongue twisters were the rule. For example, *Physalis annua ramosissima, ramis angulosis glabris, foliis dentato-serratis* was the scientific name of a herbaceous bush with round yellow berries, which is now known as *Physalis angulata*—cut-leaf ground-cherry. Linnaeus's clever system only requires two names. The first one designates the genus, and the second one, the species. In the case of seahorses, *Hippocampus* (*H.* in abbreviation) stands for the genus, followed by the name of the species—such as *abdominalis, guttulatus,* or *hippocampus.*

Despite his great services to the scientific world, Linnaeus was not infallible. He allotted the genus of seahorses to the amphibians, instead of the fish. His simple, clear taxonomic system piqued the curiosity and spurred the activities of naturalists and collectors. Scientists and hobby collectors imported dried specimens of dozens of "new" species of seahorses, among others, to Europe. The most prolific seahorse collector of the time was the Dutch physician Pieter Bleeker. In 1841, at an age of twenty-two, he signed up as a military doctor with the Dutch colonial army in Indonesia. Stationed in Batavia, today's Jakarta, he spent every free minute of his time at the local fish markets. He preserved in rum many creatures he hadn't previously known in Europe. Without access to any professional literature, he collected around twelve thousand different finned creatures. Almost two thousand of these were previously unknown to Europeans, including eight new species of seahorses.

Bleeker was not a unique case. Over one hundred different species of seahorses were said to have been discovered by researchers and naturalists in the nineteenth century alone. Some of the seahorses sighted had particularly long, narrow snouts; others had pale saddles on their backs. Seahorses from the North Sea were considered particularly similar to horses on account of their long, bristly manes. In the Mediterranean, the Suez Canal, and the Red Sea, nature lovers found pale seahorses with smooth surfaces, among others. In Southeast Asia and around the Pacific Islands, on the other hand, they found seahorses with long spines, as well as specimens with stripes like zebras, pink knobs, or with flaming red markings. So many people were enthusiastic for the rich variety of colors and shapes of these peculiar fish that biologists spoke of the "century of the seahorse."

However, the variety of species is not quite as great as seahorse aficionados believed in the past. Many of the "new descriptions" from the nineteenth century turned out to be nothing other than a few well-known species that were repeatedly "rediscovered," as has been proven by experts working with the seahorse taxonomist Sara Lourie. However, no accusations should be made against the naturalists of past centuries, because seahorses can easily confuse anyone who observes them.

In 2004, members of the international marine protection NGO called Project Seahorse (see chapters 15–17) attempted for the first time to alleviate the state of chaos in seahorse classification, eliminating the errors. Lourie, at the time a doctoral student of biology in Montreal, took upon herself the task of critically examining the species of seahorses listed in more than one hundred expert books. In the end, she was left with only thirty-three confirmed species. Yet since then, a lot has happened in the field of seahorse taxonomy—and the number of species continues to grow.

Scientists need to be particularly tenacious in this field of research. It isn't difficult for a layperson to differentiate between a gorilla and an orangutan. And on land, any horse with stripes is obviously a zebra. However, many seahorse species can only be differentiated with the help of experts. Almost all of these creatures have a similar bodily structure and, at least sometimes, many of the species look as similar as twins.

In 2006, for example, a bright orange seahorse was sighted off the south coast of England. Experts and laypeople were all excited by the fascinating new species—but it was a false alarm. Scientific examinations proved that it was a specimen of the well-known species of long-snouted seahorse (*H. guttulatus*), already discovered in 1829, which had simply changed color. The creature had probably adapted its color to camouflage itself next to an orange buoy that had sunk to the ground nearby. Most horses of the sea can change their color like chameleons, and they also seem to express their emotions via color changes. An orange-yellow seahorse, for example, could belong to almost any species that happens to be in that mood at the moment, according to British marine biologist Helen Scales.

How many different species does the genus of seahorses really comprise? Underwater photographer and fish taxonomist Rudie H. Kuiter says, with conviction, "Far more than eighty, probably even one hundred. Over thirty species of seahorses live in Australia alone." In contrast, Lourie, the most influential and strictest seahorse taxonomist in the world, recognizes less than fifty different species. She and her colleagues have gone through all the relevant professional literature, ana- lyzed over two thousand seahorse specimens in twenty-eight museums located in seven different countries, and conducted complicated genetic comparative studies. She explains the process—the identification involves excluding possible species until you have narrowed down a specimen's characteristics to match only one. It's helpful to use characteristics—such as stripes, saddles, or bands on their skin. In addition, the size can give an indication of the species. However, the length is often confusing, because a layperson can only rarely tell if the seahorse specimen is a juvenile or adult.

It's most helpful to count the number of physical charac- teristics, such as the number of bony rings around the body, or the number of fin rays that support the fins like a delicate skeleton—in addition to measuring the length of the snout and analyzing the shape of the crown or coronet. The habitat where the seahorse is found can also be useful in identifi- cation. Some species have only been found in great depths, whereas others only live in shallow water, such as mangroves.

Finally, the genome should be analyzed, which is often the primary tool now used to differentiate species. Lourie emphasizes that although the physiognomy of an individual species of seahorse is bizarre and fascinating, the basic physical structure of all the species in the genus (*Hippocampus*) is quite uniform. Their body is covered in bony plates, which are interlocked in encircling rings. This makes the bodies of almost all the species appear segmented. Most seahorses have eleven rings or ridges on their trunk and between thirty-four and forty tail rings. These structures, which are also bony, form the connections between the plates. In addition, seahorses characteristically have fifteen to nineteen fin rays on their dorsal and pectoral fins.

However, there are differences in detail that make taxonomic identification easier. Seahorses that dance to a different tune include the big-belly seahorses (*H. abdominalis*), which have one or two trunk rings more than the majority of the other seahorse species, as well as a sensational number of tail rings (forty-five to forty-eight) and fin rays (twenty-five to twenty-nine). And none can be confused with the rare bullneck seahorse (*H. minotaur*) from southeast Australia, near Tasmania, which only has eight trunk rings and seven to eleven dorsal fin rays.

The pygmy seahorses comprise a subgroup of species, including the well-known Bargibant's seahorse (*H. bargibanti*) and Denise's pygmy seahorse (*H. denise*). The unique characteristic of these tiny creatures, which are less than one inch long, is that the males don't have a brood pouch. Instead, males carry their babies in their body cavity—somewhat like female mammals. Another unusual characteristic is that pygmy seahorses only have a single gill located centrally on the back of their head. In addition, their body's external armor is less marked than other seahorses, and their appearance closely resembles their own habitat—whether it be a coral fan, seagrass, bryozoan, sea mat, or moss animal.

Most seahorses are extremely well camouflaged, making them very difficult to find in nature. Nevertheless, it's worth keeping your eyes open while diving or snorkeling, emphasizes Lourie, because it's quite possible there may still be one or another undiscovered species of seahorse drifting out in the seas somewhere.

In 1999, American hobby diver Denise Tackett, from the state of West Virginia, discovered the smallest seahorse in the world east of Borneo. She and her husband, Larry, had been diving for sponges for years, with a total of over six thousand dives. Her greatest dream was to discover new natural substances that could be used, for example, to fight cancer. Since sponges filter out infectious microbes from the water while feeding, they have to protect themselves from those microbes with antibodies. Useful bacteria that live in the sponge's walls help them with this. Scientists have already isolated over two thousand medical substances from various sponges, and perhaps someday they'll also find something to counteract malignant tumors.

The Tacketts often went diving in the tropics. In addition to her fascination with sponges, Denise Tackett had a weak spot for the tiny Bargibant's seahorse (*H. bargibanti*), a species with a bizarre appearance—bright pink or orange, with strange knobs or warts on its trunk and head that make it look almost exactly like its host coral. At a length of under an inch, these creatures were considered the smallest seahorses in the world. However, Tackett had never spied a live specimen, until one day a diver colleague told her that he often saw tiny seahorses on certain gorgonians. As a result, Tackett carefully checked each one of these fan corals and promptly found her first Bargibant's seahorse. She shot an endless series of photos. Soon, she became an expert on pygmy seahorses.

One day, Tackett and her husband were diving at a shipwreck in Indonesia, east of Borneo. On a fan coral, she saw an unusual seahorse. Tackett could tell it wasn't a Bargibant's seahorse—the creature was even tinier and more delicate looking than any she had seen, and it was slender, without the typical round belly of a Bargibant's seahorse. Instead of sedately gripping a holdfast on the coral fan with its prehensile tail, like the latter species, this seahorse was swimming around in constant motion, up and down and around the fan coral.

Soon enough, Tackett sighted more of these mysterious miniature seahorses on gorgonians of the genus *Muricella*. She was even able to witness the birth of their sea foals underwater. It had already become clear to her that this was a species that hadn't yet been scientifically described! Yet for a new species to be added to the scientific catalogue, there needs to be sufficient proof. Tackett caught a pair of the "new" seahorses. She intended to take them to the United States in a container filled with seawater in her backpack. However, her luggage was lost on her return flight—along with the seahorses. On another dive in Sulawesi, east of Borneo, Tackett caught specimens of the species again, which eventually ended up with a taxonomist at the Bishop Museum in Hawaii.

When the diver reported on her discovery in a glossy magazine, Amanda Vincent, the most famous seahorse researcher in the world, contacted her. The renowned Canadian marine biologist wanted to know everything about Tackett's discovery firsthand. She also alerted her colleague, Sara Lourie, about the potential new species. Lourie personally traveled to Sulawesi to observe the creature on-site in its natural habitat. Via telephone and the internet, they shared their findings with the experts in Hawaii, and in 2003 they drew a conclusion. The newfound seahorse, which was only 0.8 inches long, was officially deemed a new species. The researchers named it *H. denise*, after the woman who discovered it.

In the meantime, no one has doubted that Denise's seahorse represents a separate species. Over the past twenty years, many other "new" species have been discovered— more or less dubious candidates for the status.

The genetic differences between specimens of the same seahorse species are usually less than 2 percent, according to Lourie, "whereas those between species are usually 7 to 23 percent." This has been demonstrated by various research teams, and it would seem to promote clarity. However, the classification of seahorse species actually still leaves many unanswered questions, despite the Herculean work undertaken by Lourie, Kuiter, and numerous other experts.

Recently, Lourie carefully examined whether seahorse species that clearly look different could also be distinguished from one another at the genetic level. Using hundreds of tissue samples from seahorse fins, she isolated a DNA segment named cytochrome b. She was pleased to see that differences were also visible on the DNA segments of seven different seahorse morphotypes. However, seahorses of the same morphotype (i.e., with a very similar physical stature) exhibited 99 percent of the same cytochrome b genes. The Canadian taxonomist and her colleagues had found evidence supporting their hypothesis.

A little later, when other researchers conducted similar studies of tissue samples from seahorses originating in different seas, contradictions arose. The molecular biologists found an astonishingly large genetic spectrum between seahorses of the same species, which looked almost exactly the same, but were caught in different places around the world. Lined seahorses (*H. erectus*) from the Caribbean, for example, had more genetic similarities with the short-snouted seahorse (*H. hippocampus*) from Europe than with members of their own species from Brazil.

Apparently, genetics can help to better differentiate some species of seahorses; however, it would seem to be the opposite with other species. For the moment, a combination of many different factors will remain the only viable method of classifying the various seahorse species. Their appearance must be taken into consideration, as well as their courtship behavior and the structure of their DNA.

Yet does it matter exactly how many species of seahorses there are—whether it be 7, 39, or even 799 different species? British marine biologist Helen Scales says it does. Knowing the number of species in a given ecosystem, and learning how they interact with each other, is extremely informative for scientists "because tracking the various species and the sizes of their populations enables us to evaluate how ecosystems are changing, and whether they are resilient or unstable."

Over the past years, Sara Lourie has recognized two new species of seahorses. Both of them belong to the tiny pygmy seahorses. The snout of one of the newcomers looks, with a little imagination, almost like that of a piglet. Therefore, an international research team named the species *Hippocampus japapigu*—*japapigu* is the Latin name for "Japanese pig." These strange new seahorses are not the least bit shy, nor are they particularly rare. The reason they were only discovered so recently is related to their tiny size, which is only 0.6 inches. Furthermore, this tiny seahorse can easily be confused with sea tang due to its coloration. Ichthyologists at the California Academy of Sciences described the species in a scientific article; however, the scientists did not yet know very much about their newly discovered protégés. Current research has only indicated that Japanese pygmy seahorses live on soft corals or algae reefs (at a depth of up to seventy-two feet), feed on zooplankton, and are rather playful.

Above and beyond this, a species from South Africa named *H. nalu*, very similar to the *H. japapigu*, has recently been added to the scientific catalogue of seahorses. "We can currently count forty-four different seahorse species," says Lourie. In any case, it fits in well with the assessment made by seahorse expert Jorge Gomezjurado, who claimed that God might have had one too many when he created the horses of the sea.

12

My Friend the Seahorse

AQUARIUM KEEPING

"That's why we like fish in aquariums; they remind us of ourselves."

PAULO COELHO

ARE SEAHORSES SUITABLE PETS? Many people would swear to it. "Keepers give their pet seahorses names like Poseidon, Triton, Thrasher, Pace, Charlotte, Sea Biscuit, Mrs. Speckles, and Fat Albert," writes marine biologist Helen Scales in her book _Poseidon's Steed_. Others speak to their seahorses like they would to little children, singing them songs. They cry when they die. Many seahorse pet keepers believe that each of their seahorses has its own character, from lazy or shy to outgoing and quarrelsome. As a matter of fact, there are differences among them. Some seahorses will curl their tail around the finger of anyone who cleans their tank. Others are fearful and refuse their feed if it's not provided by their owner.

The keeping of pet fish began over two thousand years ago. Centuries before our times, people kept the first "house fishes" in Lydia in the southwest of present-day Turkey. These finned creatures served not only the purpose of decoration and entertainment. Flutes were played to lure them to the surface of the water, where they were questioned as oracles. If they snapped up the feed they were offered, this was interpreted as agreement. If they refused their feed, pushing it aside with their tail fin, it was interpreted as disapproval.

During the imperial Roman era (27 BCE to 476 CE), almost all the gardens of villas along the Mediterranean coastline had artificial ponds filled with salt water. Goatfish (Mullidae), colorful saltwater fish similar to bass and around twelve inches long, whose weight was paid in gold, were even kept in special ponds inside the villas. During the first century CE, the original marble walls of saltwater ponds were replaced with glass panes in Rome, Pompeii, and Herculaneum. Then people were able to carefully observe the saltwater fish underwater, instead of only seeing their shadowy movements from above.

In the meantime, pet lovers all over the world kept a countless number of species of marine creatures in aquariums in their homes—including more and more seahorses. Such tanks had their popular breakthrough in the nineteenth century in Victorian England. During that era, friends of nature wanted to learn more about the marine world, apart from legends and mythology. The great pioneer was Philip Henry Gosse from the city of Worcester.

At the age of seventeen, Gosse sailed to Newfoundland, where he worked as a merchant and collected insects as a hobby. In 1839, at twenty-nine years old, he returned to England and continued his nature studies. He became particularly interested in the sea and marine creatures. Soon he began researching fish in the wild, as well as at home in glass tanks filled with salt water. Gosse was the person who coined the term "aquarium" in 1854, in his book of the same name, where he provided helpful tips for friends of the sea to set up their own "miniature ocean." His book became a huge success. More and more people attempted to catch marine creatures and keep them in ponds or tanks. "England has become Gosse-ified, whole families at a time," proclaimed the British newspaper *The Atlas* in 1856.

Three years earlier, in May 1853, the first large-scale public aquarium was opened in London. Over four hundred species were kept in the London Fish House. Gosse caught the creatures himself on the coast of Devon and shipped them to the capital city. Seahorses were also put on public exhibit for the first time. In 1859, a certain Senhor Pinto from Portugal safely brought several seahorses from the estuary of the Tagus River near Lisbon in a goldfish bowl, after traveling for seven days—while repeatedly refreshing the salt water with a shot of oxygen via a needle. These creatures quickly attained cult status.

Several other large aquariums opened, and more and more special journals for aquarium keepers were founded, while people became ever more fascinated with the horses of the sea. When seahorses moved into the Jardin d'Acclimatation in Paris in 1866, this was even reported by the

Glasgow Herald in distant Scotland. The seahorses at the zoo in Brussels were lovingly described by a reporter from the *Daily Telegraph* in London in 1869 as fascinating creatures with "prickly manes, and a motor in the idle of their backs that strongly resembles the fly wheel of a musical box." In 1873, a British newspaper reported that in the capital city, the seahorse had become the "hero of the aquarium," as well as "the pet of the London ladies."

Naturally, the world press was highly enthusiastic when a herd of sea foals was born at the aquarium in Manchester— probably the first seahorse babies that were born out of the sea. In public aquariums in Scarborough, Brighton, and Hamburg, seahorses grew to become the favorite of the crowds in the Victorian era. However, little was known about their way of life. "They don't have any hands to shake," wrote a reporter, "so they shake each other's tails instead."

More and more people became passionate about keeping fish in the second half of the nineteenth century and acquired an aquarium. Yet these artificial ecosystems were more difficult to maintain and keep alive than Gosse described in his book. In bourgeois households, it was usually the servants who were supposed to take care of the aquariums and catch any crabs that had fled, for example. "They firmly believe they are improving their minds," wrote Wilkie Collins in his novel *The Moonstone*, of a butler complaining about the owners of an aquarium, "when the plain truth is, they are only making a mess in the house."

Nevertheless, ordinary people, as well as world-famous researchers, remained loyal to their aquariums. Among them, zoologist Konrad Lorenz, the founder of comparative behavioral science and a Nobel laureate in medicine, was a great friend of these artificial ecosystems. "A man can sit for hours before an aquarium and stare into it as into the flames of an open fire or the rushing waters of a torrent. All conscious thought is happily lost in this state of apparent vacancy, and yet in these hours of idleness, one learns essential truths about the macrocosm and the microcosm."

Even state-of-the-art technology can only mimic to a certain degree the natural habitat of seahorses and other fish in the sea. Therefore, keeping pet fish and breeding them in aquariums is not for the faint of heart. However, it's fortunate that passion has always made people inventive. "Although we didn't have any hippopotamuses," says biologist and aquarium keeper Karl-Heinz Tschiesche, recalling former times in East Germany, "the visitors drove our employees crazy." Plenty of children and adults came to see the seahorses in the aquarium, which they liked even more than the sharks, he says. "Seahorses were always among our main attractions."

The daily operation of the aquarium wasn't easy. "During the communist era, it was very costly for us to buy our seahorses," remembers Tschiesche. A pet shop in East Berlin was the only one in all of East Germany that regularly sold tropical species. Even the glass panes for the aquarium had to be acquired via complicated networks.

When procuring seahorses, the aquarium usually purchased five or six specimens at a time, recalls Tschiesche. If this was impossible, then he was able to rely on personal contacts among the crews of East German merchant ships. "We provided the crews with aquariums on board. During their leisure time, while their ships were at dock for long stretches in tropical harbors, the sailors caught sea creatures or reef fish. One sailor, who often crewed on ships to Cuba, regularly brought us seahorses that he'd caught himself. Those were the days," says the former aquarium manager with a grin.

Seahorse babies were often born in his aquarium. "Depending on the species, we often had hundreds of newborn seahorses," recalls Tschiesche. "The male expelled them out of his pouch in waves like little clouds of creatures." However, it took a long time until he and his colleagues were able to raise any of the seahorse babies. Not until a horse breeder from a stud farm started working at the aquarium did things improve. This horse breeder managed to fatten the tropical saltwater copepods, used to feed the baby seahorses, with farmed green algae, so they became more nutritious and better-tasting to the baby seahorses. He was able to raise more than thirty seahorse babies from one brood. "I think that was the first time seahorses had been bred in East Germany, to be followed by further stories of success," says Tschiesche.

And nowadays? The fascination with seahorses continues in many places. Large aquariums are highly successful in breeding seahorses, especially in the United States. At Birch Aquarium in San Diego, more than a dozen different species of seahorses have been successfully bred since the mid-1990s. Over three thousand seahorses have been sent to a total of sixty-five different public aquariums.

Progress has also been made in Europe. In Austria, for example, intensive research has been conducted on the life and nutrition of the slender seahorse (*H. reidi*), "*Hippocampus reidi* has practically become a house pet," says Daniel Abed-Navandi, deputy director at the House of the Sea in Vienna. "The life span of these creatures has increased drastically from two up to eight years in the aquarium." Other zoological aquariums are employing the Viennese methods for seahorse breeding and keeping. "We produce our own plankton nutrition chain of high-quality microalgae and microplankton," explains the biologist. "These copepods and rotifers [*Brachionus*] are fed to the newborn seahorse babies in a turbulence-free plankton-circuit aquarium, which simulates the open sea." Here the juvenile seahorses grow for several weeks until they have reached the stage where they become ground-dwellers and are slowly acclimatized to the feed that they receive in the adult stage, which primarily consists of frozen *Mysis*.

From time to time, the Viennese House of the Sea also sends its seahorses out into the world. "When the international zoo community inquired in 2014 whether we'd be able to spare a few of our creatures to send them to Asia, it seemed like quite an honor," says Abed-Navandi. Ocean Park Hong Kong was building a new seahorse tank and was looking to acquire creatures born in aquariums. Fifteen slender seahorses (*H. reidi*) were sent from Vienna to Hong Kong, packed in a plastic container with sufficient water and oxygen for the two-day trip.

The fact is that public aquariums provide good advertising for private aquariums in the home. In North America, the trend of keeping pet fish is booming and has become a popular hobby in its own right. At least 700,000 of the aquariums in the United States are saltwater, although they are more difficult to care for than freshwater aquariums.

Keeping seahorses is considered difficult and challenging because they need to feed on plankton organisms around the clock. Some species can get used to feeding on the fry of fish that are easy to breed, such as tilapia. However, a mixture of various copepods is preferable. Regarding temperature, light, and pH level of the water, seahorses are fussy. Their digestive organs are small and simple, so their tanks need frequent cleaning. Nevertheless, seahorse expert Rudie H. Kuiter encourages enthusiasts. "Other than a few species, which are particularly difficult to keep outside their habitat in the wild, seahorses can be cared for by serious aquarium keepers."

As the early pioneer of aquarium-keeping Philip Henry Gosse knew well, seahorses are susceptible to several illnesses. For example, a person should regularly check whether the seahorses in the tank are "coughing," or if they are rubbing themselves against the glass panes on account of a skin infection. If their fins are hanging down, this often indicates they are ill. If their air bladder collapses, they will sink helplessly to the bottom of the tank. And males may develop air bubbles in their pouches, making them float on the water surface like ping-pong balls.

Treating ill seahorses is a challenge, since the fish tolerate only very low doses of medication. Several specialized fish veterinarians, such as Sandra Lechleiter from Germany's Black Forest, have the required experience. "There's clearly a trend in keeping fish as pets," says the veterinarian with a smile. More and more people are establishing close relationships with their fish, she observes, "like other pet keepers have with their dog or cat." These people are prepared to spend money on veterinarian care, which is often costly. Lechleiter even makes house calls in a vicinity of up to 150 miles.

Can beginners keep seahorses in an aquarium? Or would it be better to first gather some experience in keeping fish that are less sensitive? Actually, says Elena Theys, the seahorse whisperer from Germany, you need an aquarium of around fifty-two gallons and a good source of feed. "Live feed is only necessary for pygmy seahorses or live-caught seahorses." All other species are satisfied with live feed as an occasional snack. "Really good frozen feed is important, particularly *Mysis*, which should be supplemented with vitamins from time to time," says the pet shop owner.

In addition, a good protein skimmer and holdfasts are mandatory for the horses of the sea. Many seahorses need stones, corals, or tiny caves, whereas others will even accept plastic pipes as holdfasts. "Finally, you'll need around three hours a week for feeding and cleaning," says the expert. "And that's about it." She even encourages busy pet lovers. "In contrast to popular belief, well-fed adult seahorses can also survive a weekend without feed. That should never be the rule, but an exception is okay."

Pacific Seahorse

Hippocampus ingens

13

Viagra With Fins

THE SEAHORSE
AS A REMEDY

"Water, taken in moderation,
cannot hurt anybody."

MARK TWAIN

THE CUSTOMS OFFICERS at the Brussels airport are used to lots of things—smuggled goods from diamonds to drugs to weapons. In the beginning of June 2017, they were suddenly confronted with illegal fish. Three Chinese citizens were carrying 2,063 dried seahorses in their luggage. As an ingredient for remedies used in traditional Chinese medicine (TCM), the smuggled seahorses were worth around 20,000 US dollars in total.

The three Chinese men were in transit from Sierra Leone to Beijing and had made a short stopover in Brussels. They claimed they'd been working as fishers in Africa, where they'd noticed that domestic fishers were throwing away seahorses as bycatch, even though they're considered quite valuable in China. Over the course of two years, they had collected over two thousand seahorses to share with their relatives at home. This case was taken to court. Their defense attorney argued that they were "uneducated fishermen, who knew nothing about international treaties. They had no idea they were doing anything wrong." The Belgian state prosecutor found that this was nothing but lies. The accused had traveled to Sierra Leone for the purpose of buying seahorses for the black market. Eventually, the defendants were convicted and each sentenced to fifteen months in prison, half of their time on probation.

This may sound like a bizarre once-in-a-lifetime case, yet it's actually the tip of the iceberg. The global trade in dried seahorses is a gigantic business. On June 26, 2019, the Chinese authorities confiscated 1.3 tons of smuggled seahorses in the port of Qingdao, a city on the Yellow Sea with a population of over 9 million. The estimated value of the smuggled seahorses was around 2 million US dollars. The prohibited shipment was discovered during a routine inspection of containers shipped from Peru to China, despite the fact that Peru is one of the 184 nations which have signed the Convention on International Trade in Endangered Species of Wild Fauna and Flora (CITES), officially strictly limiting the export of seahorses (see chapter 16). However, enormous quantities of dried seahorses are used in TCM, and they sell for an illegal market price of up to 1,400 US dollars per pound in Bangkok or Hong Kong—over four times higher than the current world market price for silver.

Served in soups, teas, or rice wine, dried ground seahorses are said to lengthen one's life and protect against nervous system ailments. TCM apothecaries also claim that dried seahorses will alleviate shortness of breath, pains, asthma, incontinence, digestive complaints, and arteriosclerosis. They are also used in attempts to heal broken bones, blisters, infected wounds, ulcers, liver cirrhosis, kidney insufficiency, and throat infections. They are even claimed to be a miraculous remedy for male impotence—a kind of Viagra with fins, a highly lucrative market. Over 1.5 billion people use TCM remedies around the world, according to estimates by the World Health Organization. In China, such remedies—many based on plant and animal ingredients—are used on a daily basis, like aspirin or paracetamol in the Western world.

The most important trading center for dried seahorses is Hong Kong. Millions of these creatures are bought and sold there each year. The district of Sheung Wan is considered the epicenter. Its narrow streets are crowded with delivery vans and pushcarts overloaded with boxes of dried mushrooms, herbs, and berries—as well as seahorses. Often, the seahorses are piled up high, nested like spoons, in plastic crates.

TCM apothecaries think seahorses are a valuable raw ingredient. Proponents of TCM point to a thousand years of tradition. The mythological founder of TCM, the "divine farmer" Shennong, was said to have collected herbs in the mountains of central China, testing their healing powers on himself. His name is found on the first book of TCM, the *Shennong bencaojing*. However, historians have discovered that it was learned people in the Han dynasty (206 BCE–220 CE) who wrote the work, more than three thousand years after the death of the mythological founder of TCM. Plants, mushrooms, worms, snakes, and centipedes are classified in it as remedies, but seahorses are not!

Half a millennium later, in the year 793 CE, seahorses were mentioned the very first time in a Chinese book on herbal remedies. In it, they are described as "sweet and warm," and recommended for fortification of the kidneys. This characterization is also found in a basic volume on TCM from the sixteenth century CE, the *Bencao cangmu*, which is considered the magnum opus of the Chinese medic Li Shizhen. He included almost two thousand natural remedies in his book, with seahorses among them. "Sweet tasting," he too noted about the horses of the sea, as well as their positive effect on the function of the kidneys. In addition, seahorses were said to relieve purulent skin infections, as well as to help facilitate childbirth.

Is this nothing but a load of superstitions? Believers in TCM claim it's holistic. They're convinced that a mysterious life energy named qi regulates the human body and the entire universe in a continuous cycle of life, transformation, and death. At the same time, the flow of qi ensures the harmonious coordination of the complementary principles of yin and yang. If the flow of energy and the elements in the body are out of order, then a person becomes ill.

This therapy claims to eliminate the roots of the illness. Holistic healing is the goal, not just the treatment of symptoms. An individually mixed remedy of plant and animal ingredients—such as powdered seahorse—is supposed to support the healing process. An increasing number of people believe in this practice.

Nevertheless, TCM was already considered a thing of the past around one hundred years ago. In 1929, Mao Zedong commanded the Chinese Red Army to "eradicate all shamanic traditions and superstitions." Yet while the communists were taking power, Mao realized how many Chinese people strongly believed in TCM—and that there was no practical alternative to the (formerly cheap) natural remedies, especially for the millions of poor people in the country. Pharmaceutical medications from the West were only available in large metropolitan areas at the time and were not affordable for most Chinese people.

In 1949, Mao gave in and officially changed his policy. He announced that TCM "is a great treasury. We must do everything to continue unraveling its secrets." As a result, there was a revival of this kind of traditional healing. In the 1960s, the communist leaders recruited so-called barefoot doctors. They held crash courses teaching the basics of Western medicine alongside the main corpus of TCM—and sent the graduates into rural areas of China. These folks wandered from village to village to improve medical care for the farmers. More than a million barefoot doctors are still working in the vast rural areas of China. Among other remedies, they're still treating their patients with dried, powdered seahorse.

Yet what good could seahorse tissue really do as a remedy? In studies, very low levels of the male sexual hormone testosterone and the amidosulfuric acid taurine were found in the bodies of yellow seahorses (*H. kuda*), as well as in bulls' testicles. A modern TCM handbook also refers to animal experiments in which female mice were fed with an alcoholic extract of seahorses. As a result, the ovaries of these rodents were said to have grown at a faster rate than those of their compatriots, who weren't fed this special concoction. The seahorse extract even miraculously extended the sexual cycle of the female mice. Yet does any of this make it more likely that dried and pulverized seahorse tissue can strengthen the male member, or work as an antidote for asthma, or stimulate hair growth?

British scientists at the University of Plymouth have researched over one thousand studies conducted around the world investigating the healing potential of TCM remedies. Only three of those scientific analyses were conducted as randomized double-blind studies—as required by evidence-based Western medicine—during which neither the doctor nor the patient knew whether they were receiving the test remedy or a placebo. One of the double-blind studies demonstrated that TCM remedies were not able to reduce the side effects of chemotherapy for cancer any better than a placebo. The results of the other two double-blind studies did not allow any clear conclusions to be made.

However, perhaps the question of potent individual ingredients is misleading. Due to the holistic approach of TCM, proponents argue that the efficacy of their remedies is generally much more difficult to prove empirically than that of a tablet for headaches, which is only intended to alleviate that type of pain. According to the traditional teachings of TCM, it is not about curing individual symptoms, but healing the whole person.

During the research for her book *Poseidon's Steed*, marine biologist and science writer Helen Scales also took a close look at TCM. Seahorse remedies, if one follows TCM, are particularly beneficial for kidney function, relates Scales. According to Western medicine, the role of this pair of organs is to filter the blood and eliminate waste products via urine. That overlaps with Chinese medicine to some extent, according to Scales, which is why seahorses are used to treat incontinence in Asia.

According to TCM, the kidneys also take on other important functions. The so-called qi, or life force, absorbed via the lungs, flows along a meridian directly into this pair of organs, where it is stored. From the kidneys, it then flows into other parts of the body. Traditional medicine in China teaches that if the kidneys are too weak, they're not able to access the qi from the lungs. Based on this belief, it's not particularly surprising that dried seahorses are also supposed to indirectly help relieve shortness of breath or asthma by strengthening the kidneys, elaborates Scales.

The British science writer claims that it might also at least be plausible that seahorse extracts could aid in allaying impotence, if a person believes in such energy meridians. According to TCM, sex is an exchange of yin and yang. The concept is that during sexual intercourse, the woman gives the man yin, and the man gives the woman yang. If a man does not have sufficient yang, he won't be a good lover, according to this principle, elaborates Scales—whereby TCM attempts to allay the problem with dried, pulverized seahorses.

It is indisputable that animal rights lovers heaved a great sigh of relief when the synthetically manufactured potency pill known as Viagra was marketed in 1998. Thanks to this drug from the pharmaceutical industry, men would hopefully stop using tiger penises, rhinoceros horns, seal genitals, and seahorses for the purpose of strengthening their erections. Sadly, that wasn't the case. Even the scandalous test results of researchers at the end of the 1990s—proving that TCM remedies contained extremely high concentrations of lead (lead has been scientifically proven to impair male potency!)— did not result in a lower demand for dried seahorses.

In addition to TCM, Japanese Kampo medicine also believes in seahorse remedies for the purpose of improving the libido. The traditional jamu medicine of the Indigenous peoples of Indonesia also use horses of the sea to treat impotence, as well as rheumatism and loss of memory.

In Europe, seahorses have also undergone a long career as a healing remedy. During ancient times, horses of the sea didn't play a role in mythology alone (see chapter 5). Pedanius Dioscorides, an early pioneer of Western pharmacology, who studied in Tarsus and Alexandria, published a teaching compendium by the name of *De materia medica* between 50 and 70 CE, which also lists remedies including seahorses. This basic compendium became well known all over the Roman Empire and remained influential in Europe for over 1,500 years. In it, Dioscorides recommends the ashes of a burnt seahorse mixed with goose fat massaged into the scalp to alleviate hair loss. Furthermore, a mixture of manatee lard, honey, and seahorse ashes was used as a remedy to cure leprosy.

As a result, further gray eminences prescribed seahorse remedies. The scholar Pliny the Elder recommended yellow seahorse in alcohol for prostate complaints, and black seahorse in alcohol for impotence. The philosopher Claudius Aelianus, on the other hand, advocated the use of pulverized seahorse mixed with vinegar and honey for the treatment of dog bites at the beginning of the third century CE.

This tradition continued until early modern times. In the Renaissance era, the doctor and naturalist Conrad Gessner of Zurich swore that the tissue of seahorses was a miraculous remedy for poor vision, hair loss, side stiches, and rabies, as well as for a weak libido or incontinence. "This creature, worn on a chain on one's body, shall motivate one to intimate relations. Item dried, pulverized, and ingested, shall wondrously help against bites from rabid dogs," he writes. Furthermore, "this creature burnt to ashes, when mixed with strong vinegar essence, fills the bald head with hair once more. The powder of the dried horse of the sea alleviates side stitches, and taken with a meal, aids those who cannot withhold their urine." Even in the year 1753, the noble readership of the British *Gentleman's Magazine* could read that genteel ladies in Italy made use of a seahorse elixir to improve the quality of their maternal milk.

Not until the nineteenth century did natural medical remedies become less popular in Europe. Scientific proof that infections were caused by pathogenic germs finally became common knowledge, and synthetically manufactured ingredients as antidotes for such pathogens became widely used. Only a few select natural remedies of plant or animal origin remained common.

Today, seahorses are no longer used as remedies in Western medicine. However, the demand for them in the rest of the world is greater than ever. Around eighty countries are currently involved in catching, trading, and consuming seahorses, according to research done by the marine conservation organization Project Seahorse. Around 20 million seahorses end up in international trade each year, mainly for use in TCM. Global marketing via the internet has generated a huge problem. More and more seahorses are being captured, ground to powder, mixed with other substances, and sold in the form of pills.

Yet there really isn't any good reason for this! In 2006, a scientific study in the United States surveyed 145 doctors who treated their patients according to the principles of TCM, with regard to the significance of seahorses in their treatments. Almost two-thirds of the doctors confirmed that seahorses have hardly any relevance whatsoever for the healing properties of the remedies they prescribe. Some of these doctors also confirmed that seahorses can easily be substituted with *Epimedium* flowers, walnuts, or human placenta. Only 2 of the 145 doctors surveyed in North America—around 1.4 percent—listed seahorses as being an important ingredient in their remedies.

Myth has it that pulverized seahorse is also able to extend penis length; interested men would likely do better for themselves by purchasing a sports car and leaving the world's seahorses alone.

14

Learning From Seahorses

ROBOTICS & CO.

═══

"I think the biggest innovations of the twenty-first century will be at the intersection of biology and technology. A new era is beginning."
STEVE JOBS

HUMANITY IS PRODUCING huge volumes of garbage—even in outer space. In low orbits alone, up to a height of six hundred miles, more than 6 million pounds of space debris is flying around our Earth, including pieces of rockets, wreckage from explosions, old satellites, screwdrivers, as well as garbage bags from the Russian space station Mir containing frozen fecal matter, among other dangerous waste. Over 22,000 pieces of space junk with a diameter of at least four inches are traveling through space at a speed of up to 17,000 miles per hour.

Now, Mother Nature—or bionics, to be exact—is being called upon to help clean up the mess we've made. Scientists and engineers want to make use of natural phenomena for the purposes of technical innovation. Researchers from the Swiss Space Center at the Swiss Federal Institute of Technology in Lausanne were inspired by the tentacles of sea anemones in designing prehensile arms for flying robots, intended to help collect the space debris. In overcoming the problems we've created on Earth, humanity is also taking a cue from marine creatures' boxes of tricks.

Some researchers are investigating the bony structures of fish fins, for example. If you press your finger against fin rays, they don't bend away; they bend toward your finger instead. Engineers would like to use the fin-ray effect for the construction of flexible building roofs, such as for swimming pools (so the roof can be opened during good weather, transforming an indoor pool into an open-air pool).

Seahorses have served as a model for designers and engineers in several forms. Around the turn of the millennium, for example, experts from Japan investigated which shape of pillow was most conducive to helping commuters take a nap on the train. They tested dozens of pillow shapes. The best results were achieved by a sixteen-inch-thick pillow in the shape of a seahorse. Yet the development of these pillows didn't make anyone rich. Japanese train companies decided against making any wholesale purchases—because it would be embarrassing for men to be seen holding onto a seahorse in public, according to the train company's PR agent.

Nevertheless, it's quite possible that the seahorse's prehensile tail will be the key to success in other research projects. Despite its robust bony plates, that organ can not only be rolled up into a frontward spiral, but also (to a certain degree) backward or sideways. This enormous flexibility is achieved because the bony plates, connected along the entire length of the tail via filigree joints, can glide over one another. In addition, seahorse tails have a special muscular structure only found in fishes: a combination of back muscles running parallel between connective tissue structures, enabling fast grasping, and highly efficient ventral muscles that can maintain the grip over lengthy periods of time.

The bony armor itself serves as a buffer against impacts and vibrations. Experiments have shown that the bony plates of the seahorse's exoskeleton can be pressed together to around one-half of their size without breaking. It's no wonder that materials researchers are being inspired by the biochemistry of the seahorse's armor in developing flexible, yet robust, synthetic materials.

Robotics researchers, on the other hand, are learning from the design of the seahorse's tail. Most seahorses' tails have around thirty-six segments that are connected by joints. "Normal" prehensile tails, such as those of monkeys, squirrels, or macaques have a round or oval-shaped cross section. Seahorses' tails are unique, and their long, self-rejuvenating tails have a rectangular cross section. Michael Porter, professor of engineering at Clemson University in South Carolina, asked himself why. To investigate, he made a three-dimensional printout of a seahorse tail in plastic and compared its strength to that of a similar construction with a round cross section.

When the researchers tested the two models by contorting them, bending them, and striking them with a rubber hammer, they determined that the rectangular tail was more elastic than the round one, and it also could withstand more pressure. The angular bones didn't splinter or break, even if the tail was compressed to 60 percent of its regular size. Round bones, on the other hand, clearly proved less stable. The rectangular tail makes it more difficult for predators to bite through a seahorse.

This wasn't the only advantage of the rectangular tail shape that came to light in the scope of the study, which Porter and his colleagues published in the journal *Science* in 2015. When gripping, the rectangular tail provided a greater contact surface than the round model—enabling, for example, a better grip on a metal rod. If the rectangular tail was twisted, it automatically returned to its original position—without expending any further energy—as soon as the pressure was released. Researchers on Porter's team hypothesize that this mechanism might also protect the seahorse's sensitive spinal cord when it's attacked by predators.

"This study demonstrates that engineering designs are convenient means to answer elusive biological questions when biological data are nonexistent or difficult to obtain," the scientists said.

In addition, they would also like to learn how to develop new technologies from nature, as the seahorse's prehensile tail is a jackpot for robotics. "Engineers tend toward building things that are inflexible, so they can control them more easily," explains robotics researcher Ross Hatton from Oregon State University, who collaborated on Porter's research project. "But nature makes things just strong enough not to break, and then flexible enough to do a wide range of tasks.

That's why we can learn a lot from animals that will inspire the next generations of robotics."

Based on the model of the seahorse's prehensile tail, for example, future robots will be built with artificial hands that are simultaneously highly dexterous and extremely robust. Some engineers even envision search-and-rescue robots, inspired by seahorses' prehensile tails, which crawl along the ground and self-contract to enter narrow spaces. Novel drill shafts for petroleum exploration are to be developed, as well as exoskeletons for people with physical disabilities, surgical technology for minimally invasive surgery, and prehensile arms for collecting space debris. Porter says that this special armor will enable future technologies to be simultaneously lighter and more flexible than traditional, inflexible robots, as well as more robust, with better suspension than "soft robots." All based on the principles of the seahorse's prehensile tail.

This concept is also trending in Asia. Japanese designers have decided to use it for a novel application—an extension of the human body. However, they were certainly not thinking of penis enlargement, as claimed by the so-called seahorse remedies—instead the idea is to extend the tailbone. A team working with designer Junichi Nabeshima from Keio University in Tokyo has developed a robotic tail by the name of Arque, which can be attached to the hips with a kind of belt.

This invention was presented in public for the first time at the SIGGRAPH 2019 Exhibition in Los Angeles. The artificial tail looks somewhat peculiar when it's attached to the body and wags back and forth around the height of the lumbar vertebrae. Yet the Arque is supposed to work miracles.

Nabeshima explains in a video promoting his invention that it is meant to help maintain balance like a pendulum. The twenty-eight-inch-long prosthesis consists of twelve artificial vertebrae connected with each other via joints and powered by artificial muscles. The muscle contractions are generated via an air compression system with an external compressor. The development of a battery-powered version will soon be on the drafting table.

Many animals use their tails to maintain their balance and dynamic equilibrium, according to the elaborations of the Arque developers. Therefore, a robotic tail should also be able to help human beings maintain their balance—for example, when having to move heavy furniture, or climb high stairs. Or it could give elderly people added safety, in place of a walking stick or a rolling walker.

Each of the twelve vertebrae in Arque weighs three ounces and can be made heavier by adding metal weights. In total, the balancing aid weighs up to 5.5 pounds. Whenever people wearing it move their upper body toward the left, the artificial tail will move to the right, and vice versa.

Perhaps the human tailbone, an evolutionary remnant, will not remain a stub forever. If things go according to this inventor's plans, the artificial tail will soon be used in virtual reality. As a gadget in video games, it could give the players whole-body feedback in combination with virtual-reality goggles, thus helping them improve their performance. Arque is only a prototype at present. However, Nabeshima's team of designers are already planning to build their seahorse-inspired robotic tail in various lengths and are working on a version to be launched on the market.

One of their approaches sounds convincing. Arque could be used as an extension of exoskeletons that are already being commonly deployed in heavy industry. Ever since paraplegics were able to learn to walk again with the help of similar robotic systems, this field of research has been trending. According to the designers in Tokyo, Arque will give this research field additional impetus.

In Germany, on the other side of the world, many companies have already started using exoskeletons for their employees, since musculoskeletal injuries are the most common cause of sick leave and early pension in all jobs involving strenuous physical labor. For example, Volkswagen uses several different exoskeletons in its assembly halls, according to a report in the German edition of *Technology Review* from the summer of 2019. The model for assembly work done in a squatting position—for instance, assembling a cockpit—can be attached like a backpack, and it supports and strengthens upper-body motion. Workers who do heavy lifting use an exoskeleton robot, attached at a lower point of their body, to relieve their lumbar vertebrae.

Many other companies in the automotive industry, as well as in logistics and caregiving, are testing supportive robotic technologies, according to *Technology Review*. Experience has proven encouraging. The pain and exhaustion caused by hard physical labor can apparently be reduced in many cases.

If the designers from Tokyo are right, exoskeletons will soon include an Arque application to improve balance. Learning from seahorses will mean succeeding at sea, on land, and perhaps even someday in outer space—in the fight against space debris.

Short-Snouted Seahorse
Hippocampus hippocampus

15

Terminal Trawl?

THE HUMAN THREAT

*"What we once used as weapons of war,
we now use as weapons against fish."*

SYLVIA EARLE

N DJIFFER, a fishing village in southern Senegal, seahorses are in great demand at the market. Toward the evening, brightly painted fishing cutters are beached on the sandy seashore to sell their catch. A multitude of merchants compete for the best fish—barracudas, stingrays, oysters, sea cucumbers, perch, and seahorses. The prices for seahorses, in particular, are skyrocketing, according to research by the German radio broadcaster Deutsche Welle. All over West Africa, the volume of seahorses bought and sold has increased over the past years. In fact, 600,000 of the creatures were exported from this region in 2018, according to the marine conservation agency Project Seahorse, mostly to China.

First the crews of large, deep-sea fishing vessels in Asia began drying all the seahorses they netted to sell them. Soon,

Senegalese fishers and other West African fishing boats followed suit and began hawking seahorses in the harbor of Dakar to dealers, who sold them to Asia. In the meantime, all over West Africa, traditional fishers with small boats began targeting seahorses as their catch.

The global trade in wild-caught seahorses has become a multibillion-dollar business. Almost one-quarter of all the smuggled marine products confiscated at airports between 2009 and 2017 were seahorses. Individual passengers at airports have been caught with up to twenty thousand dried seahorses, at a value of over 10 US dollars apiece.

Canadian professor of marine biology Amanda Vincent, from the University of British Columbia, had already published evidence of the horrific dimensions of the international trade in seahorses, based on a study conducted twenty-five years ago, including the dire consequences. In some regions of the Philippines, the seahorse populations had already shrunk by up to 70 percent over just two decades, and in other regions of Asia, the United States, and Brazil, Vincent determined similar tendencies.

However, the world of seahorses was still in a comparably good state two and a half decades ago. Today, around eighty countries actively trade in seahorses on all continents of the world, except Antarctica. Experts estimate that around 20 million seahorses are bought and sold each year—if not far more!

In their role as victims, seahorses are "absolutely unsurpassed," wrote science journalist Michael Miersch with a dash of sarcasm, in an essay for the magazine *Mare* at the turn of the millennium. Furthermore, "two culprits can be charged for the crime: capitalism and globalization." Miersch meant that ironically. In fact, however, it was capitalism that made some people wealthy enough to be able to afford the—now high-priced—traditional Chinese medicine (TCM) remedies made of seahorses. Next to Microsoft, Coca-Cola, and Nokia, the international conglomerate Tong Ren Tang—which sells TCM remedies—is one of the richest companies in China. Globalization catapulted trade between TCM apothecaries in China and countless Chinese expats around the world into unprecedented dimensions. TCM is no longer only sought after by diasporic Chinese populations in New York, London, or Paris. Globalization has also awakened the interest of other people in North America and Europe to avail themselves of supposedly "gentle" TCM remedies from Asia.

Yet what's the origin of the millions of seahorses used in these remedies and sold on global markets? For some divers specialized in diving for seahorses in the Philippines and Thailand, these creatures have become the bread and butter of their existence. Although a lot of seahorses are caught by these divers, the danger of being killed as bycatch in the dragnets of large trawlers is much greater—seahorses have become victims that weren't primarily targeted for catch.

The widespread exploitation is horrific. Each year, more than 50 percent of the continental shelves of the world are plundered at least once by "bottom trawling"—in marine regions that are less than five hundred feet deep, where all the known seahorse species live. That means a total area of 6 million square miles, around 1.5 times the size of the entire continent of Europe! The so-called trawl boom, a heavy-gauge boom weighted down with iron weights, is dragged along the seabed on skids or runners with open nets that tear out the kelp, seaweed, sea fans, coral, and sponges. All the large fish, as well as all the smaller fish, are caught in the fine mesh of the nets. Even huge pieces of live coral reefs are broken off and dragged along the seabed. It is an extremely brutal method of fishing—akin to hunting squirrels with military tanks on land, while killing everything along the way that isn't fast enough to escape.

Up to 73 percent of all fish and crustaceans caught by trawlers are bycatch, or unintended catch. Shrimp fishing is particularly inefficient. For each pound of shrimp caught, there are up to ten pounds of sponges, crabs, shellfish, seahorses, and other fish unintentionally caught in the nets as bycatch. Shrimp fishing with bottom trawlers is particularly fatal for seahorses, because they often share their habitat with the shrimp that are in such high culinary demand. A bottom trawler only catches around one seahorse a day now, according to random surveys conducted by Project Seahorse. However, the gigantic number of bottom trawlers around the world has an annual bycatch of 40 million seahorses—double the number of seahorses that are sold on global markets, according to experts' estimates! Even if some of these creatures are released back into the sea, many of them are so badly injured that they can no longer survive.

The saddest fact about the seahorses is that even if bottom trawling were to be prohibited around the world today, the survival of seahorses would still be uncertain, because their habitats are shrinking at such a fast pace—whether it be mangroves, seagrass beds, coral reefs, or regions of soft corals. More and more of the seahorse's habitats are being destroyed by bad fishing practices and other human destruction, such as draining bays and estuaries. Shrimp farms, yacht harbors, agricultural fields, hotel complexes, shopping centers, condominiums, and airports are all being built in place of natural seahorse habitats.

Climate change is also threatening the horses of the sea. "Global warming can lead to a fast increase in sea levels," explains biologist Ralf Schneider from the GEOMAR Helmholtz Centre for Ocean Research Kiel. "Habitats of many seahorse species, such as coral reefs and seagrass beds, may then receive insufficient sunlight and die back as a result. Seagrass, in particular, requires sunlight." Increasing silt and turbidity of the water, due to an excess of nutrients and heavy phytoplankton growth, are also fatal for these ecosystems.

The main reason for the overfertilization of the seas is industrialized agriculture. In the United States alone, large-scale livestock farming (pigs and cattle, etc.) produces over a billion tons of manure each year! The total area of agricultural fields is not large enough to utilize or dispose of such gigantic volumes of organic fertilizer. The livestock manure either ends up as dangerous methane gas in the Earth's atmosphere, or the liquid slurry seeps into the groundwater and drains into the rivers and seas, where the excess fertilizers destroy the clarity of the water.

Soft corals, the habitat of Bargibant's (*H. bargibanti*) and Denise's pygmy seahorses (*H. denise*), are even more highly endangered because of climate change, explains

Schneider. "Due to an increase in sea temperatures, dinoflagellates [symbiotic, single-celled organisms living in corals]
lose their capacity for photosynthesis and begin producing
toxic substances instead. Two or three degrees Fahrenheit
of warming suffices to cause this adverse reaction. The coral
reefs then repel the dinoflagellates, making them lose their
color, which is known as fatal 'coral bleaching.'" The corals
can die as a result, says Schneider. "And that means many
seahorses are losing their habitats."

The mangroves around the world, which have been full of
life for millions of years, are also seriously endangered. Since
the middle of the twentieth century, around 30 percent of
the global mangroves have been clear-cut—and even up to
75 percent in the Philippines. Mangroves are not only an exis-
tentially essential habitat for seahorses, but also for thousands
of other fish species and crustaceans. The small fry of fish and
young marine creatures are protected from their predators in
mangroves while they grow, before moving into the seas and
coral reefs at an adult stage. Mangroves are also indispens-
able in preventing the erosion of coastal regions, and they
even provide protection against hurricanes and tsunamis.

What about seagrass? Around one-third of all seagrass
beds have disappeared globally since the nineteenth century.
Beginning in 1990, the rate of destruction of this crucial
habitat for innumerable species has increased exponentially.
The estuaries of New Zealand, for example, have lost almost
all of their natural seagrass since the 1960s. Many marine
creatures normally spawn in seagrass beds. Seagrass emits
oxygen into the surrounding water and into the seabed via

its roots. This aerates the seabed and makes it a viable habitat for crabs, shellfish, sea urchins, and marine worms. These form an important basis of the food chains in the oceans of the world. In addition, seagrass filters waste materials from the water and forms sediments. Seagrasses also help fight climate change, because they store a huge volume of carbon dioxide. However, the excess nutrients from fertilizer runoff in the Baltic Sea, for example, have dramatically decreased the formerly vast expanses of seagrass. A fungal disease in the 1930s also decimated the seagrass beds in the North Sea, including the seahorses living in that habitat. Without seagrass, most seahorses cannot find an essential holdfast. Being forced to navigate stormy seas with their tiny, delicate fins can deplete their energy.

Successfully fighting for the survival of seahorses urgently requires dedicated work in three areas, according to the experts: radical reduction of bottom trawling, prohibition of trade in dried and endangered seahorses, and wide-scale protection of the global seas from pollution and destruction, particularly in coastal waters.

Time is running out! Experts predict that fifteen species of seahorses are already threatened with extinction. "And seahorses are like an early warning system," says British marine biologist Helen Scales, "just like the canaries that miners used to take down the mine shafts." Those birds react most sensitively to methane gas and carbon monoxide. As long as the canary continued singing and hopping around in its cage, the miners could rest assured that there was still enough clean air without toxic gases. A dead canary signaled that the mine shaft should immediately be evacuated. With regard to our oceans, seahorses are likewise a reliable indicator, writes Scales. "They are sensitive creatures that can only survive in intact habitats with good water quality." If we human beings do not manage to protect our seas, then seahorses might well become extinct, which "may be a warning of a greater ecological catastrophe in the future."

= 16 =

To the Refuge!

PROTECTION

═══════

"To cherish what remains of [the Earth],
and to foster its renewal, is our only legitimate hope."
WENDELL BERRY

THE PEOPLE LIVING in the fishing village Handumon on Jandayan Island in the central Philippines were desperate, even though they had an excellent basis from which to earn their living—catching seahorses. Not far from the coast and their wooden-stilted houses, there are coral reefs, seagrass beds, and mangroves—a paradise for seahorses.

In the 1960s, traders showed up on Jandayan Island. They wanted to buy seahorses for use in traditional Chinese medicine (TCM). Children from Handumon began picking seahorses out of the seagrass, and soon the adults wanted a part of the income. They went out night-diving for seahorses in sea canoes, illuminated by small kerosene lanterns. In the darkness, the eye of the nocturnal tiger tail seahorse

(*H. comes*) could be seen flashing in the black water. Soon, catching seahorses in Handumon became an important source of income, and almost half of the nine hundred residents of the village made their living by selling the creatures.

Other fishing villages on Jandayan Island followed suit and also began night-fishing for seahorses, although their prevalence was already decreasing. While the divers had been able to catch up to one hundred seahorses each night in the 1970s, they were satisfied with around twenty per night in the 1990s. It wasn't rare for them to return to their village empty-handed.

The sad story of Jandayan Island became known around the world as far as Canada. In 1993, an international research team led by Vancouver-based marine biologist Amanda Vincent traveled to Handumon. Not only had the seahorses in that region been badly decimated, according to the scientists, but the seahorses' habitats were also in grave danger. The research trip to Handumon became the incentive to found Project Seahorse, the largest and most important marine protection NGO fighting to save seahorses around the world.

The marine life in the vicinity of Jandayan Island was particularly endangered, due to a highly destructive fishing technique that used dynamite—often homemade devices built with artificial fertilizer. The detonations killed all the fish, small fry, and marine creatures in a radius of dozens of feet. The dead fish floated to the surface of the sea, and it was a killing for the reckless fishers scooping them up. The seabed became riddled with craters and destroyed coral reefs.

However, the fishers eventually listened to some of the concerns of the environmentalists. It became clear to them that without any seahorses, their major source of income would disappear. They agreed to measure and weigh the seahorses they caught before selling them, as well as to regularly monitor the conditions of the coral reefs and seagrass in their fishing territories. In addition, homemade cages for pregnant male seahorses were deployed. Pregnant seahorses were kept on the seafloor in cages made of fish nets strung over wooden frames until they gave birth. The newborn seahorses were tiny enough to easily slip through the mesh of the nets. However, the male seahorses remained in the cages and were dried and sold after they had given birth.

Nevertheless, the fishers often had disputes concerning who was allowed to sell the male seahorses. Eventually, the cages for pregnant seahorses were discontinued. The marine conservationists of Project Seahorse came up with a new strategy. In coordination with the local fishing community, they dedicated a marine protection refuge of eighty-one acres (about the size of sixty-one football fields) north of Handumon in 1995. Fishing and catching of seahorses was temporarily prohibited there. All of the seahorses in the marine refuge were registered and numbered. Custodians with diving goggles and snorkels regularly monitored the population to be sure none of the seahorses went missing. A patrol boat with people from the fishing village was responsible for ensuring that the fishing prohibition was adhered to at all times. It was a great success. Withing a few years, the populations of many seahorse species in the marine refuge had recovered.

Soon, further coastal waters were closed for fishing boats. After just ten years, Project Seahorse's highly impressive program had designated twenty-two marine protection refuges, each with an area between 25 and 125 acres in size. The populations of many fish species in these protected zones recovered, and random sampling indicated that an increasing number of seahorses could be found north of Jandayan Island—a good sign. However, monitoring surveys later indicated that the situation was very troubling. The coral reefs in the region had still not recovered in 2010 from the destructive dynamite fishing, and the populations of seahorses, the raison d'être of Project Seahorse, had not increased to any significant extent.

Wasn't the project on Jandayan Island successful? "That's a complicated story," says marine biologist Sarah Foster of the University of British Columbia in Vancouver, who also monitors international trade in seahorses and works for Project Seahorse. "The Philippines belong to the countries that have generally prohibited catching and trading seahorses since 2002, in keeping with the CITES agreement," she says. However, the night fishers from Handumon either lost their means of income or they were forced into illegal fishing practices as a result. "The good news is that there has just been a change in the fishery laws and regulations in the Philippines, which has brought them into harmony with the CITES regulations." Currently the Philippine government is being advised on the development of a "management plan for seahorses," to enable the trade of these highly sought-after fish in a legal and sustainable framework.

Without doubt, regional approaches alone will not suffice to save the world's seahorses. The international trade in these creatures urgently needs to be strictly limited—and there were good signs this would take place at the beginning of the twenty-first century. An article in the German magazine *Der Spiegel* in November 2002 said, "For seahorses and turtles, there is new hope." The Convention on International Trade in Endangered Species of Wild Fauna and Flora (CITES) had just convened in Santiago, Chile, and several new species had been listed as endangered, including seahorses—all species of seahorses, since customs officers around the world cannot be expected to differentiate between each of the many species. Yet what became of this ray of hope?

In the meantime, 184 countries have signed the CITES agreement. The convention's honorable goal is to prohibit any international trade in all species that are listed as endangered, unless it is "not damaging" to the local animal populations in the individual countries. The problem is that even experts can hardly estimate how many seahorses of a certain species live in the marine waters of any given country. Researchers of the University of Bologna launched a seahorse population count. They recruited 2,500 hobby divers and had them survey the numbers of seahorses along the 4,700-mile-long Italian coastline. The project ran for three years. During a total of around six thousand hours of underwater observation, the divers found over three thousand seahorses. However, no differentiation was made between the species, or whether individual specimens had been counted more than once. So, how can a serious estimate be made for the number of seahorses that would be ecologically sustainable to catch?

For pragmatic purposes, the marine protection environmentalists switched to a simple rule of thumb on a global basis—the size of the seahorses. Only seahorses with a minimum length of four inches may be caught and traded. This limit, which might seem highly arbitrary due to the difference in size between dwarf and giant seahorses, is nevertheless reasonable. Specimens over this size limit have almost always reached adulthood and have already had a chance to reproduce.

The night fishers and seahorse divers in the central Philippines generally adhered to this size limitation and released any smaller seahorses, allowing them to survive and reproduce. Moreover, alternative sources of income were generated on Jandayan Island in cooperation with Project Seahorse: ecotourism with seahorse watching, as well as the cultivation of seagrass as a natural fertilizer.

Indonesia, India, Malaysia, the Philippines, and Thailand even prohibited any export of seahorses in accordance with the CITES agreement. However, what had seemed like a great success in the protection of marine life almost proved the opposite instead. Not only were seahorses still being caught and sold in those countries that signed the convention, but also a booming international black market developed that was almost impossible to regulate. At the beginning of 2019, a group of scientists working under the auspices of Canadian marine biologist Sarah Foster for Project Seahorse determined that the CITES agreement had not reduced the trade in seahorses. Foster and her colleagues interviewed 220 merchants in Hong Kong on the origin of the seahorses they were selling. Based on their interviews, they were able to calculate that up to 95 percent of the seahorses were being imported from countries with an official prohibition on the export of seahorses! "Dried seahorses can easily be smuggled over borders," says Foster. "They are small and keep for a long time. Smugglers often hide them in their personal luggage."

Just like Amanda Vincent, the great pioneer in the protection of seahorses, Foster also thinks that a general prohibition in the trade of seahorses will not succeed. People should not have their source of income taken from them; instead, sustainable fishing and trading of seahorses according to CITES regulations and criteria should be permitted. "This would also provide an incentive for improved sustainable fishery management, which should also have a positive effect on the protection of wild seahorses."

However, there is a further problem. CITES does not have the legal mandate to impose any sanctions. The marine protection environmentalists can only appeal to the goodwill of decision makers in those countries that have signed the convention. They often encounter people who ignore their appeals. "CITES needs to work more closely with the countries that signed the convention to ensure that intermediary traders do not buy seahorses from countries where trade in them is prohibited, and the authorities must impose high fines," Foster insists in no uncertain terms. Over the past years, the authorities in Hong Kong have at least attempted to restrict illegal imports and are threatening smugglers with prison sentences. In 2018 alone, forty-five shipments of dried seahorses, weighing approximately 1,036 pounds, were confiscated—in total, around 175,000 seahorses that were supposed to be pulverized for dubious use in TCM remedies.

"The main problem for seahorses is not their use in traditional Chinese medicine," says Foster, "it's the fishing industry." Fishing fleets should be reduced, and bottom trawlers should be kept far away from marine protection zones with drastic consequences for any infringements, urges the seahorse expert. Over 95 percent of seahorses sold on global markets are bycatch, mostly from bottom trawlers fishing for shrimp. "Project Seahorse is campaigning against bottom trawling, which arbitrarily exploits seahorses and other marine creatures, thereby threatening to destroy the biological diversity and sustainability of the marine food chain," says Foster. "For this purpose, we need the help of all seahorse fans—to ensure that this urgent issue can no longer be ignored by governments everywhere!"

The marine biologists and environmental protection activists of Project Seahorse are not going to give up. In the meantime, the NGO has already become active in thirty-four countries on six continents. More than thirty-five marine protection zones have been designated in the Philippines, Vietnam, and Hong Kong, usually in coordination with local marine protection organizations. In many of these protection zones, the fish populations have increased significantly over the past years.

On a political level, a lot has changed in connection with the protection of the seas over the last few years, emphasizes Foster. In 2008, not even 1 percent of the global seas was designated as a marine protection zone; in the meantime, around 8 percent of the seas now have some form of protection. However, some researchers remain unsure about this approach. They believe that marine protection zones only serve greedy fishing industries and governments as a fig leaf to cover up the increasingly brutal exploitation of all the other marine regions.

The British marine biologist Callum Roberts is of a somewhat different opinion. "Prohibiting overfishing in certain regions has repeatedly proven to be a highly effective means of protecting marine life," states the professor at the University of Exeter. Yet a piecemeal approach will not suffice. "If marine protection zones are to be effective, there must be many of them and they must be interconnected in a broadscoped, well-monitored network. We cannot simply monitor a few individual sites of spectacular beauty. Instead, we need to use interconnected marine protection refuges as the new foundation for all our marine activities."

Wouldn't it be feasible to replenish the populations of marine creatures with fish bred in aquariums and released into the wild? The experts are skeptical. Experiments conducted by researchers in Australia have shown that not enough juvenile fish can be raised in captivity to competitively cover the huge demand for use of seahorses in TCM. Dried seahorses used by TCM apothecaries are usually not sold per piece, but by the pound, or even by the ton.

A few seahorse aquaculture fisheries, particularly in Asia and the United States, are financially successful nevertheless. Their targeted markets are seahorse aquarium keepers, prepared to spend good money for their hobby. After all, hundreds of thousands of seahorses are bought as pets each year. Creatures from aquaculture are better suited for keeping in aquariums. And only a few wild-caught specimens still live in zoos and in public aquariums. This serves to protect the wild populations to some degree.

In contrast, there is controversy among experts about releasing seahorses raised in aquariums into the wild. Zoologists are debating whether seahorses from aquaculture are even capable of living in the wild. In aquaculture, inbreeding often takes place, making these creatures less resilient for life in the open seas.

The Australian marine biologist David Harasti from Port Stephens Fisheries Institute near Sydney has taken a more refined approach to support the wild populations. Among other things, he temporarily feeds wild seahorses in an aquarium. In 2007, he caught several adult White's seahorses (*H. whitei*) in the harbor and transferred them to a saltwater tank. The seahorses mated there, had babies, and at an age of four months, he released the juveniles. His goal was to increase the survival rate of juvenile seahorses. In the wild, baby seahorses are constantly in danger, and often only one in two hundred juveniles survives to adulthood.

White's seahorse, which can grow up to six inches long and changes its color according to its mood, needs special protection. A large population of the species had always lived in the seagrass beds between Sydney and Forster, 180 miles to the north. However, practically their entire habitat was destroyed by devastating storms and building development between 2010 and 2013. Huge volumes of sand buried the soft corals, sponges, and seagrass in their habitat. Ninety percent of the local seahorse population was eradicated. It was White's seahorse, which has been red-listed in the catalogue of endangered species by the IUCN (International Union for Conservation of Nature) since 2017, that inspired Canadian marine biologist Amanda Vincent. She was so entranced by the mating rituals of this species in the 1990s that she became the world's leading expert in the fight to protect seahorses.

Harasti soon realized that feeding juvenile seahorses in aquariums would no longer suffice to ensure the survival of White's seahorses. He came up with a novel idea. Since their natural habitats had been so badly damaged, maybe he could construct a habitat for them himself—a kind of retreat or refuge for seahorses. Eventually, an entire chain of "hotels" was set up in Sydney Harbour. Fortunately, the investment costs were low, since Harasti wasn't designing luxury retreats.

He had often observed groups of White's seahorses on abandoned lobster traps while diving. Therefore, he built the first prototype of a hotel according to the model of a lobster trap—a ten-by-ten-foot cage with a frame of metal bars and walls of wire. The marine biologist hoped that algae, sponges, and corals would populate the cages within a short period, creating a "natural" habitat for seahorses. Copepods, brine shrimp, and other zooplankton would also be attracted—the perfect prey for horses of the sea.

When the biologist and his team set up the first seahorse hotels in the spring of 2018 in Sydney Harbour, the target group reacted skeptically. However, over the course of two months, the creatures became less timid. By the end of the year, sixty-five White's seahorses checked into the hotels, as well as countless other fish, octopuses, carpet sharks, algae, corals, and sponges. It was a great success when the first males became pregnant.

In 2019, Harasti and his team opened nine more seahorse hotels in Sydney Harbour. Transferring the pregnant sea-horses to the SEA LIFE Sydney Aquarium, for the purpose of giving birth and caring for the babies, proved advantageous. In April 2020, they were again able to release dozens of seahorse babies born there to the hotels in Sydney Harbour. The Australian concept became a model. In Greece, Indonesia, the Philippines, the United States, and Portugal, experiments are being conducted with similar seahorse refuges.

Harasti still speaks of "hotels" and not of homes. "Because I had only planned for the seahorses to live there for a limited period of time," he says. On the other hand, he and his team noticed that many of the horses of the sea had developed a strong attachment to "their" refuge. They were repeat-edly observed in one and the same place during monthly monitoring. One seahorse remained in the same hotel for more than an entire year—and that was just the beginning.

In the meantime, Harasti had even developed a personal relationship with some of the seahorses in his hotels. "Dawn" was his favorite seahorse for a long time—a brilliantly yellow-colored female seahorse that he had observed at least five years previously for the first time while diving. He says that Dawn became locally famous. Four years ago, she met a dark gray seahorse stallion in the hotel that Harasti named "Dusk." They became a dream couple. The biologist thought they would probably remain in their romantic hotel for a lifetime, after he had observed them there for over three years. How-ever, that wasn't to be. In January 2020, Dusk disappeared completely. Dawn was alone in the hotel. She seems to have taken it bravely, the marine biologist says. Maybe Dusk only left to get himself a snack at a nearby convenience store? (Sadly, plenty of garbage from convenience food wrappers is floating in the sea.)

To put it cynically, if Dusk wasn't consumed by a predator, at least he was living proof that Harasti's hotels were conceived as temporary housing and not as prisons. Unfortunately, this model might become the future for seahorses everywhere, not only around Sydney. Artificially constructed oases for seahorses, surrounded by empty seas, dead zones, and concrete.

Harasti is optimistic it won't be that drastic. He says that before he began experimenting with the hotels, many people in Sydney had no idea that seahorses lived there at all. In the meantime, people have become highly enthusiastic about the hotels. He and his team hope that the hotel project in Australia has not only raised awareness and interest in the endangered White's seahorse, but also in many other marine creatures. Other rare species, such as nurse sharks and sea dragons, are also suffering on account of environmental pollution, heavy shipping traffic, and loss of habitat on the east coast of Australia.

At least it appears that the strategy of raising awareness about marine life with the help of seahorses is highly successful down under. David Harasti's seahorse hotels have become hotspots for hobby divers, who hope to observe seahorses in the sea for the first time.

Lined Seahorse

Hippocampus erectus

= 17 =

Workhorses

SEAHORSE POWER TO END THE CRISIS?

---

*"The great use of a life is to
spend it for something that outlasts it."*

WILLIAM JAMES

BEFORE UNTYING the dock lines, the crew members are getting some new body art. A tattoo artist from Amsterdam has been hired especially for this purpose. Henk Schiffmacher, also known as Hanky Panky, has also tattooed rock stars, such as Kurt Cobain and Motörhead frontman Lemmy. It is May 2011, and he is tattooing Greenpeace activists with a smiling seahorse—the logo for the environmental organization's Cleaver Bank campaign.

Shortly afterward, the campaign ship leaves the dock. A rainbow-colored seahorse flag is flying from the mast, also designed by Hanky Panky. The cruise destination is the Cleaver Bank sandbank, ninety-nine miles off the coastline of the Netherlands. Greenpeace activists regularly climb smokestacks or chain themselves to train tracks. However, in May 2011, they're on a mission of a different kind. They'll be releasing seahorses into the sea.

With the help of a special crane, the activists will be placing seven-foot-tall seahorse sculptures on the seafloor. It's not an art school for fish; instead, the goal is the protection of marine life. Cleaver Bank was long known for its great diversity of sea creatures. Countless species of fish lived near the sandbank, as well as crabs, anemones, and cold-water corals. However, fishing boats had badly exploited the delicate ecosystem with bottom trawling.

In 2007, the Dutch government declared Cleaver Bank an environmental protection zone. Yet bottom trawling, the highly destructive fishing technique that completely destroys all marine life on the seafloor, was not prohibited. That proved fatal for seahorses, in particular. Not only were innumerable seahorses caught in the dragnets, but their holdfasts of seagrass and coral were also decimated.

How to rectify the fatal situation? The Greenpeace activists came up with a concept based on seahorse power. The homemade sculptures of seahorses that were lowered onto the seafloor with the help of cranes had three purposes. First of all, in the course of the rescue campaign, they held a symbolic power and were eye-catchers for the press. Second, they provided artificial underwater holdfasts for live seahorses. And above all, the sculptures were made of solid wood with a granite base weighing several tons—robust enough to keep all bottom trawlers away from Cleaver Bank.

The Greenpeace activists were well aware that dragnets from trawlers were plundering all the fish in the sea—and these huge, heavy sculptures acted as obstacles that would keep away the enormous dragnet booms.

Greenpeace of the Netherlands notified the local fishing trawlers about the positions of the seahorse sculptures and sent the coordinates to the Dutch coast guard. The goal of the campaign was not to damage the bottom trawlers' dragnets, but to prevent them from further plundering and destroying marine life—with plenty of seahorse power! A fourth prong was added to the first three in 2011, when the seahorse became the mascot of the campaign, explains biologist and Greenpeace project head Pavel Klinckhamers. "The former Dutch minister of economics, agriculture, and nature was an enthusiastic horse breeder. Since he cared for horses, we thought he might also care about seahorses."

All in all, the Greenpeace activists' campaign was successful, although—or maybe even because—the local fishing trawlers actually heaved some of the troublesome obstacles back out of the water. "They were trying to prove that we had created a dangerous situation for them," says Klinckhamers. "Those fishers even took the case to court, hiring one of the most famous lawyers in the Netherlands, but they lost in court nevertheless."

"The fight for seahorses continues," says Klinckhamers. "While we're speaking, our British colleagues, farther north on Dogger Bank, are also placing rocks from cliffs on the seafloor as obstacles against bottom trawling," he explains. Those rocks aren't decorated with wooden seahorses, but the seahorses at Cleaver Bank seem to have broken the ice for similar protection campaigns in other countries to combat the plundering and exploitation of the seas.

That would be entirely in keeping with the goals of the great pioneer of seahorse protection. In 1996, Canadian biologist Amanda Vincent was already highly concerned about the survival of a lot more than just the charming seahorses. She considered seahorses a kind of mascot for the protection of all endangered marine life. In the meantime, the scientist and environmental lobbyist has found many like-minded colleagues around the world.

In 2005, for example, marine biologists in Asia noted with great concern that more and more mangroves and seagrass beds were being destroyed on the southwest coast of the Malay Peninsula. The people in those coastal regions seemed indifferent to the destruction. However, when studies demonstrated that the estuary of the Pulai River was not only the largest seagrass region in all of Malaysia, but also had the only significant population of yellow seahorses (*H. kuda*), environmental activists saw a chance to raise awareness among the locals. They founded the marine protection organization SOS Malaysia—standing for "Save Our Seahorses." Although Malaysia would like to follow Singapore's path to economic prosperity, seahorses should not have to pay the price.

In the estuary of the Pulai River, north of Singapore, the container port of Tanjung Pelepas is continually expanding, which endangers the local population of yellow seahorses more than ever. Even the seahorse protection activists of SOS

Malaysia weren't able to stop the expansion of the container port. However, the activists were able to raise the awareness of many Malaysian people concerning the environmental problems. Volunteers from far inland repeatedly came to the coastline to lobby for the horses of the sea.

While other seahorse friends were building refuges for seahorses or lowering seahorse sculptures to the seafloor to protect the seahorses' habitat, Adam Lim, the head of SOS Malaysia and a scientist working for the Institute of Ocean and Earth Sciences (IOES) at the University of Malaya, began catching dozens of seahorses with landing nets and transferring them to a safer habitat, along with his team.

Much farther north, Lim had already found another hotspot—coastal waters where pipefish, rare crocodiles, and fork-tailed dugongs (similar to manatees) were living. "In Malaysia, there are thirteen different species of seahorses," he says, with shining eyes, including five different kinds of pygmy seahorses, tiger tail seahorses, three-spot seahorses, and hedgehog seahorses. He wants to promote ecotourism and offer excursions to watch seahorses, glow worms, and rare seabirds. That also has great economic potential, he emphasizes.

"Over the past ten years, we have marked around 850 seahorses," says Lim with pride. Very few environmental protection organizations in the world can make such a claim. SOS Malaysia wants to document the development of these populations in detail. Lim is sure of one thing: if seahorses could, they'd have made an SOS call long ago, in Malaysia and many other countries. In the summer of 2020, he published a crowdfunded seahorse book for children. It should raise awareness and generate enthusiasm for these wonderful creatures in the next generation—as well as help children become engaged in protecting the seas and marine life.

In Europe, seahorses are also becoming a figurehead for the protection of marine life and the seas. Recently, António Pina, mayor of the small Portuguese city of Olhão, decided to erect a large seahorse monument. "Seahorses are part of the heritage of our city," he declared, and designated the seahorse the mascot of Olhão. "They will bring our citizens good luck, but above all, they should warn us against the further destruction of our coastal waters."

A seahorse, alongside a pipefish, also serves as the logo of the current campaign of the German environmental protection organization BUND against overfertilization of the seas. The huge volume of pig and cattle manure that ends up in the seas and rivers via agricultural runoff has led to overfertilization of the seawater. This directly results in the extreme growth of phytoplankton and other fast-growing, problematic algae, which badly cloud the seawater. The subsequent lack of sunlight kills the plants on the seafloor, such as seagrass, tang, or slow-growing macroalgae—and the marine ecosystem dies out when its equilibrium is lost.

Due to its low rate of seawater circulation, the Baltic Sea is particularly threatened. There were once huge expanses of seagrass growing at a depth of up to one hundred feet. Today, not a single stalk of seagrass can be found at a depth of more than twenty-three feet, due to the severe clouding of the seawater from overfertilization. When the seagrass beds and tang forests died out, the habitats and nurseries of many marine creatures and their young were obliterated, especially those of pipefish and seahorses.

Yet perhaps the BUND campaign—with its charming logo—has already made a difference? Over the past few years, the total area of seagrass in the North Sea and the Baltic Sea has increased slightly. The overfertilization caused by intensive livestock farming has been reduced a little, and the seagrass beds in the North Sea and the Baltic Sea are slowly recovering—so that even seahorses have been observed again from time to time along the German coastline. A tiny ray of hope.

"The global fight for the survival of seahorses and healthy seas must continue," says Pavel Klinckhamers, who is currently working on a new Greenpeace campaign in Taiwan. The environmental activists of Project Seahorse, the organization founded by Amanda Vincent, have not lost any of their enthusiasm. Project Seahorse is training the next generation of scientists, environmentalists, and activists who are campaigning to protect our oceans. According to their annual report:

> Seahorses are flagship species for a wide range of marine conservation issues. As with many marine species, they are threatened by harmful fishing practices, such as bottom trawling, as well as habitat degradation. Project Seahorse is dedicated to ensuring that seahorse populations and their habitats are healthy and well-managed.

The NGO's campaigns to protect seahorses also help protect thousands of other species: "Saving the seahorses means saving the seas!"

If everything goes according to the environmentalists' plan, the curious world of seahorses will survive—helping to prevent the endangered marine ecosystems of our oceans from collapsing. A tall order for such tiny creatures!

Of Giants and Dwarfs

A GUIDE TO THE COOLEST HORSES OF THE SEA

"To know that you are unique makes you beautiful."

HARNAAZ SANDHU

World Champions of Camouflage

In 1969, the same year humans landed on the moon, the first **BARGIBANT'S SEAHORSE** (*H. bargibanti*) was sighted. The tiny creature was desperately clinging to a soft coral, which New Caledonian marine biologist Georges Bargibant—after whom the seahorse was later named—had fished out of the water to install in his aquarium. He didn't notice the seahorse until it was in his tank. No wonder, since the tiny creature's gray body is covered with bright pink or orange-yellow knobs, which are exactly like the buds of the coral it lives on. Regarding the creature's ability to camouflage itself,

Bargibant's seahorses, only one inch long, eclipse all other species. They can imitate the color and shape of certain fan corals so well that even experienced ichthyologists—if they ever manage to find Bargibant's seahorses—think they have found several different species.

There are two different color morphs of these seahorses: purple with pink or red knobs, adapted to the finely branched fan coral *Muricella plectana*; and yellow with orange knobs living on *Muricella paraplectana* fan corals. However, all Bargibant's seahorses seem to prefer the former species of coral as the best place for mating.

Typical characteristics of this seahorse include a crown consisting of a rounded knob instead of points. On the right and left sides of its face, it has an elongated knob over the eye, as well as rounded cheek tubercles. Its body and trunk appear fleshy. Rings on its trunk, typical for many other seahorses, are barely visible. Bargibant's seahorse has a markedly short snout, and its dorsal fin is stronger than that of most other seahorses.

Male and female Bargibant's seahorses look almost exactly the same. The only difference is that the males have a vertical slit on their belly—the opening of the brood pouch where they carry their baby seahorses. The females have a lump in this area, which upon closer examination can be identified as the ovipositor, a phallic-like organ which transfers the egg cells to the male's brood pouch.

American deep-sea diver Denise Tackett has observed that these seahorses usually live on fan corals in bonded pairs. However, it's also possible for several pairs to form a community on a fan coral. Once Tackett even sighted fourteen pairs on one gorgonian—including a few single Bargibant's seahorses. In general, it can be said that the population density of this species is extremely low—often only one specimen in a surface area of six thousand square feet.

In 1999, Denise Tackett and Sara Lourie were likely the first divers in the world to witness a Bargibant's seahorse giving birth to baby seahorses. All together, it took around fifteen minutes for thirty-four babies to be born, each around two millimeters long. That seems to be about average. In aquariums, Bargibant's seahorses also usually give birth to around thirty-five seahorses per brood.

The Promiscuous Ones

DENISE'S PYGMY SEAHORSES (*H. denise*) are more slender than Bargibant's seahorses, yet they otherwise look rather similar. Their orange-colored body is decorated with dark rings on the tail. They don't have any spines or a pronounced crown. The sexes of this species can be relatively easily determined; the mares are larger and thinner than the stallions.

Denise's pygmy seahorses are more flexible in their choice of host coral in comparison to their cousins, the Bargibant's. However, just like the Bargibant's, they have optimally adapted their color and body shape to their host coral, making them particularly difficult to find. The species was named after the American diver Denise Tackett, who discovered them around twenty years ago in the Indonesian sea near Sulawesi (see chapter 11).

Denise's pygmy seahorses live more closely together than Bargibant's seahorses. They are mostly active during the day, spending their time hunting and feeding. Social contact and mating usually take place in the early morning or during sunset.

The pregnancy of Denise's pygmy seahorses is unusually short, and the number of offspring per brood is low. After only eleven to fourteen days, around seven to fourteen seahorse babies are born. Nevertheless, these seahorses seem to enjoy their sex lives. Once, divers observed a Denise's pygmy seahorse giving birth to thirteen babies within four minutes—only to mate again just fourteen minutes later.

Denise's pygmy seahorses are livelier and more dynamic than Bargibant's seahorses. They often leave their host coral, roll up their tail, and swim around the gorgonian. They are also more sociable. Some specimens have intimate relations with several partners. One female of this species was observed conducting a threesome with two males—the only proven case of polyamory in the realm of seahorses. Both males only received half a brood of eggs, and the female was able to minimize the risk of a failed pregnancy—by not putting all her eggs in one basket, so to speak.

The Nano Fraction

SATOMI'S PYGMY SEAHORSE (*H. satomiae*) was named after the Japanese diver Satomi Onishi, who discovered it on the coast of Borneo in 2003. These horses of the sea only grow to half an inch in length, making them the smallest seahorses in the world—so tiny that a single specimen could fit in the space of a human thumbnail. These seahorses are mainly differentiated from all the other pygmy seahorses—the group of particularly small, thinly armored species with only one gill on the back of their head—due to the conspicuous spines on their head and body, which give them a martial appearance. Further typical characteristics are a dark spot by the eyes, white spots distributed all over the body, and an angular coronet. When diving at night, you might spy entire groups of Satomi's pygmy seahorses, usually in the vicinity of gorgonians. During the day, they are rarely observed. They live in depths of thirty to sixty feet. Their breeding season is in the fall; the newborn seahorses are black and around three millimeters in size. Above and beyond this, even experts hardly know any further details about this tiny, secretive creature.

The Rotund Giants

BIG-BELLY SEAHORSES (*H. abdominalis*)—yellow, light brown, or white, and with spots often on body, head, and tail—are the largest seahorses in the world. From the tip of their crown to their tail, they can measure up to fourteen inches long. Like the tiny pygmy seahorses, they have twelve (or sometimes thirteen) trunk rings. In addition, they have the longest dorsal fins, and more black rings than any other seahorse species. A further typical characteristic is the conspicuous filament-like spines on their head, which help camouflage the seahorse—often among water plants. Big-belly seahorses continue growing throughout their lives;

however, the older they are, the slower they grow. In an aquarium, they can live to an age of eight years old.

The physical appearance varies greatly in this species, particularly the length of the snout. Some experts, such as Rudie H. Kuiter, distinguish a separate species—*H. bleekeri*, which has a particularly long snout. In the case of this (potentially) separate species, a female ready to spawn usually attracts several males, who attempt to seduce her to give them her eggs by distending their brood pouches. *H. bleekeri* seahorses are easy to breed and well-loved in aquariums around the world. However, other experts, such as taxonomist Sara Lourie, doubt whether these seahorses really constitute a separate species—and they have good reason. Both short-snouted and long-snouted specimens are found in each brood of big-belly seahorses.

These seahorses can become pregnant at any time of the year. Their offspring swim in an almost horizontal position, not curling up their prehensile tail until they become adults. This peculiarity was unknown for a long time, so that juvenile big-belly seahorses used to be falsely classified as a separate species.

The rotund giants among the seahorses also have the greatest physiognomic differences between male and female. In comparison to the females, the males are heavier, with longer prehensile tails—as well as shorter, wider snouts.

The corpulent giants are more active during the night than the daytime. Divers have observed these fish coming together in schools late in the day. Yet it still isn't clear what purpose this may serve. They often use macroalgae as holdfasts, as well as nets, piers, or flotsam drifting in the sea. As a rule, big-belly seahorses live in rather shallow coastal waters, preferably in protected habitats, such as bays. Once, a specimen of this species was found at a depth of 341 feet—extremely deep for seahorses. Apparently, these giant horses of the sea are full of surprises.

Champion Boxers

SHORT-HEADED SEAHORSES (*H. breviceps*) are closely related to the big-belly seahorses. However, they only grow to around two to four inches in length and have a column-like coronet, as well as a short snout. Their yellowish-brown to reddish-violet bodies are covered in conspicuous white spots with dark borders, particularly on their heads. The front side of their tail has light horizontal stripes. Short-headed seahorses often have a sort of mane on their head and neck, formed of skin filaments. The filaments or spines on these creatures are of varying length; some are short and pointed, while others are rounded like knobs.

One theory on the evolution of this species hypothesizes that over the course of millions of years, the smallest big-belly seahorses (*H. abdominalis*) mated with each other, while the larger ones mated among themselves. Therefore, today's short-headed seahorses were the great-grandchildren of the smallest big-belly seahorses many millennia ago. They live along the coast of Australia—preferably in rocky reefs and habitats characterized by algae, but also in seagrass.

Behavioral scientists have discovered that short-headed seahorses—atypical for horses of the sea—do not form bonded pairs; rather, they live in mixed social groups of up to ten individuals. During the entire courtship and breeding season, they remain in such schools. In order to make themselves more attractive for sexual mates, the male short-headed seahorses fight among themselves. They wrestle with their prehensile tails and "box" each other with their heads. The loser flees, and the winner has a good chance of finding a female willing to mate. Scientists have observed intensive social interaction of single males with up to three females. During one study spanning five weeks, each female only responded to the advances of a single male. Perhaps short-headed seahorses actually do lead monogamous lives, like most other horses of the sea, despite the dramatic hierarchic battles between the stallions (see chapter 8).

Night Owls

TIGER TAIL SEAHORSES (*H. comes*) were named after their characteristic yellow and black striped prehensile tail. They grow up to seven inches in length and were among the first tropical seahorses observed on-site in their natural habitat. They have a low crown with five rounded points. Most tiger tail seahorses live as pairs in reefs with plenty of soft corals, as a rule at depths of over sixty-five feet. They live everywhere along the coastline of Southeast Asia and are most common in the Philippines.

Atypically for seahorses, this species is primarily active at night. During the daytime, tiger tail seahorses usually hide in coral reefs. Not until sunset do they emerge from their hiding places to cling on to their holdfasts, usually one and the same coral branch, where they ambush their prey. Normally, a single pair shares a territory. Tiger tail seahorses appear to be highly monogamous and remain true to their territory and holdfast. Researchers have observed the same two specimens at the same place over almost two years during the course of a study.

This species can become pregnant throughout the entire year; however, the main breeding season is from September to November. The pregnancy takes from thirteen to twenty days; however, it is shorter in the warmer season than in winter. The birth can take several days, during which around four hundred baby seahorses are born.

Since the 1960s, specialized fishers have been catching tiger tail seahorses in the Philippines. Despite the official government prohibition on catching or trading in seahorses as a result of the CITES agreement of 2002, these creatures are unfortunately still being caught and exported (see chapter 16).

Amanda's Favorites

WHITE'S SEAHORSES (*H. whitei*) count among the midsized horses of the sea and grow up to six inches long. These gray-brown or yellow-colored seahorses were named for the UK surgeon general John White, who first published an image of the creature in a journal in 1790. White's seahorses have high crowns or coronets and live on the east coast of Australia, particularly in the region around Sydney. This was the first horse of the sea to be scientifically studied over an extended period of time. The great pioneer in this field of research, Canadian professor of marine biology Amanda Vincent, observed and analyzed their social and sexual behavior in the 1980s. She conducted most of her research in Watsons Bay in Sydney Harbour, where a large colony of these creatures was living. The behavioral scientist spent hundreds of hours underwater. She discovered that White's seahorses live in close monogamous relationships, somewhat similar to matrimony. Vincent also observed that the courtship of this species can take over an hour, and that changes in their colors take place while they dance together in the underwater world, stallion and mare often entwining their tails. While diving, she also observed the pair rising together toward the surface of the water, the female transferring her eggs, and several male pregnancies. As a staunch feminist, the seahorse researcher was particularly fascinated by the unusual attribution of the sexual roles in seahorses (see chapter 9), of which most zoologists around the world were not yet aware at that time.

White's seahorses are found at depths of up to thirty-nine feet in seagrass beds, and on soft corals and sponges. Individual adults have also been found in large estuaries during the winter months. In this species, most pairs remain together and have offspring several times a season. Around sixty to seventy baby seahorses are born in each brood. In contrast to most other foals of the sea, they do not have a pelagic stage. Instead, the babies immediately sink to the floor of the sea (see chapter 4).

The habitats of White's seahorses near Sydney were badly destroyed by storms and building development between 2010 and 2013, and the local population of these creatures has been decimated. Since 2017, the White's seahorse has been on the red list of endangered species of the International Union for Conservation of Nature (IUCN). In the meantime, Australia has taken special measures to protect these seahorses—for example, some juvenile creatures are being temporarily housed and fed in the SEA LIFE Sydney Aquarium (see chapter 16).

Severely Overfished

THREE-SPOT SEAHORSES (*H. trimaculatus*) have been named after the three conspicuous dark spots on their back—although not all members of the species have these typical markings. Some three-spot seahorses are brown, others are spotted, a few are golden orange, while others are black-and-white striped, or pitch black. Typical characteristics are the clearly visible hooklike spines on either side of their neck and above their eyes, the narrow head, a very long prehensile tail, and a very low coronet.

Three-spot seahorses are widely distributed. They live along northern Australia, as well as in Tahiti, the Philippines, Japan, and India. Their habitat constitutes sandy, open areas of the seafloor at depths from 30 to 320 feet, and they grow up to six inches long. The males are usually somewhat larger than the females.

Their predatory behavior conforms with the mainstream of seahorses. They feed on zooplankton and ambush the tiny creatures while clinging on to seagrass stalks to save their energy. Scientists assume this species is one that holds on to algae rafts and drifts far distances at sea to populate new habitats. A few years ago, three-spot seahorses were observed for the first time in the Strait of Malacca, the sound between the Malay Peninsula and the northeast coast of Sumatra, in very shallow waters less than three feet deep, which is highly unusual for this species. Those were probably the first three-spot seahorses ever to be observed by scientists in this region.

A huge number of three-spot seahorses have been pulled out of the sea in the dragnets of bottom trawlers. Of all species, these horses of the sea are the most sought-after as a remedy in traditional Chinese medicine (TCM). Experts estimate that this species alone constitutes one-third of all the seahorses traded worldwide (see chapter 13).

Endangered Southerners

KNYSNA SEAHORSES (*H. capensis*) are the only seahorses that can survive in almost pure fresh water. Due to their special metabolism, they are capable of switching between seawater with a salt content of 3.5 percent and almost pure fresh water (with only 0.1 percent salt content).

Another of their special characteristics is that Knysna seahorses have the smallest territory of all horses of the sea. The region they live in only comprises nineteen square miles—they live exclusively in the estuaries of South Africa. Researchers assume they are among the rarest seahorses in the world. According to estimates in 2008, there were only around 900,000 members of this species. Today, there may be as few as 1,000. All live along the famous Garden Route between Cape Town and Gqeberha (formerly Port Elizabeth), not far from the coastal city of Knysna—which was an important harbor for trade in gold and ivory in the nineteenth century and is primarily known as a tourist destination today.

Knysna seahorses are severely threatened. The meteorological caprices along the Western Cape can be extreme. In 1991, after a strong downpour, three thousand of these creatures were washed ashore, where they suffocated. Moreover, their tiny habitat is shrinking further and further. In this densely populated coastal region of South Africa, more and more hotels, golf courses, and housing complexes are being built. Sewage is draining into the sea, and motorboats are destroying the seagrass. The Knysna seahorse was therefore the first seahorse species to be listed as endangered by the International Union for Conservation of Nature (IUCN) more than ten years ago. In the meantime, the South African Fisheries Act prohibits catching these seahorses, and part of the Knysna Lagoon has become a dedicated marine protection area.

Typical for Knysna seahorses is a smooth body almost free of spines, a rounded head without a coronet, and a short, stumpy snout. (Juveniles may have a small coronet, which disappears as they grow.) Knysna seahorses are usually brown with dark flecks—but there are also white, yellow, orange, beige, green, and black specimens. At over five inches long, they count among the midsized seahorses. Like many other horses of the sea, they live in bonded pairs. They reach sexual maturity at an age of just under one year. The Knysna seahorse was discovered around the year 1900.

Classics Without Spines

YELLOW, COMMON, OR SPOTTED SEAHORSES (*H. kuda*)—their scientific name was taken from the Malaysian word *kuda* meaning "horse"—not only live in estuaries, but also in mangroves or seagrass habitats that aren't located in river estuaries. At up to seven inches in length, they count among the midsized seahorses. The males are often not yellow at all, but black with fine vertical lines of white spots, or eggshell color with brown spots. The crown of this seahorse is tilted backward; its snout is broad. Normally, yellow seahorses live at depths of up to 108 feet. According to estimates, they are one of the most common seahorses in the world, as their name suggests. However, almost all spineless seahorses in the Indo-Pacific were considered to be members of this species for a long time.

This species is also popular in traditional Chinese medicine (TCM). Dealers often bleach specimens in the sunshine to get higher prices, since colorless seahorses are considered to constitute a better remedy than dark ones. Yellow, common, or spotted seahorses are also one of the favorite horses of the sea for marine aquariums. There are fish breeding enterprises for this species in China, India, and Vietnam.

In western Malaysia, the marine protection organization Save Our Seahorses Malaysia (SOS Malaysia) is responsible for protecting the yellow seahorse. The container port of Tanjung Pelepas is vastly expanding in the estuary of the Pulai River north of Singapore, endangering the local population of seahorses. Volunteers working with marine biologist and environmental activist Adam Lim catch the seahorses with landing nets in shallow water and organize their relocation to a safe coastal region a few miles away (see chapter 17).

Bullnecks

H. minotaur is known as the **BULLNECK SEAHORSE**, and it can grow up to two inches long. It has a larger than usual head, a thick neck and trunk, and a more delicate tail. Another special characteristic is that these seahorses lack a clear segmentation between their head and tail, and they have a very short snout. Both their head and their trunk appear unusually fleshy. They have neither body rings nor spines nor any other particular decorative feature.

Bullneck seahorses are eggshell-colored and speckled with brown spots. Their scientific name is reminiscent of the minotaur of ancient Greek mythology—a fabled creature with a huge bull's head and a human body. They have a tiny dorsal fin with only seven fin rays. They are lacking bony plates on the ventral side, and most of these creatures only have a slight lump or mound on their head, instead of a coronet. All three known specimens of this species were caught by fishing trawlers off the southeast coast of Australia, at a depth of 210 to 360 feet. Their spineless, smooth body leads experts to hypothesize that this seahorse lives on sponges.

Beauties With Tiny Babies

SLENDER SEAHORSES (*H. reidi*) live in the Caribbean and along the Brazilian coastline, where they are most common in estuaries. Their favorite holdfasts are mangrove roots or driftwood floating in the tide. A study has shown that there are more females of this species than males, and the mares are more dynamic and active than the stallions. Without a doubt, the slender seahorse is one of the most fertile of all seahorses. Up to 1,600 babies are born in each brood, and the juveniles become fertile at an age of only two months. The babies can hardly be seen with the human eye, making them the tiniest offspring of all seahorses—although the adults can grow up to seven inches long and are among the largest horses of the sea.

Slender seahorses are relatively thin. They are mostly smooth-surfaced, but they also have a strange, convoluted coronet. They are well-loved as pet fish due to their conspicuous coloring—from brown to bright orange, or yellow and red to black. Some specimens have a light-colored pattern on their back, reminiscent of a saddle. They are bred for aquariums, and specialized divers also catch them in Brazil, at a depth of up to twenty-three feet. Brazil is one of the largest exporters of pet fish, and slender seahorses are also used for ceremonies in certain Brazilian religions, such as Candomblé.

The taxonomic status of several local populations remains unclarified, such as that of the so-called Brazilian giants of uncertain origin, which marine pet shop owner Elena Theys has kept and bred in the past (see chapter 1). Perhaps those seahorses were a distinct species, or maybe they are only a subspecies of *H. reidi*.

Slender seahorses are closely related to West African seahorses (*H. algiricus*). Researchers assume that members of both species can even mate with each other, if they have the opportunity. However, even Sara Lourie, the strict taxonomist, still classifies them as different species.

Europeans With Manes

LONG-SNOUTED SEAHORSES (*H. guttulatus*) are usually dark green or brown, and decorated with wavy lines of tiny white dots. The French naturalist Georges Cuvier, one of the founders of modern zoology, was the first to describe this species in 1829. They really do have a long snout—at least in comparison to the other European seahorse, named *H. hippocampus*. Further characteristics are the long spiny growths on their head and back, which with a little imagination look like a mane. This helps the creature camouflage itself in seagrass beds.

Long-snouted seahorses prefer to hide among seagrass stalks, sea urchins, or bryozoans. In general, they like habitats with dense growth on the seafloor, and their holdfasts are any kind of water plant, invertebrate, or artificial structure. They're among the species that use floating plants in the sea as a ferry—that could explain their relatively widespread distribution. They live in the North Sea as far north as Scotland, and in other regions of the northeastern Atlantic, as well as the Mediterranean and the Black Sea.

They have a higher population density than that of the short-snouted seahorse (*H. hippocampus*). Studies in Italy and Portugal have shown that around twelve times as many long-snouted seahorses were sighted there than short-snouted seahorses. The long-snouted seahorses at home in Europe mate between April and October. Males gather in groups around females ready to spawn, and they compete to mate with them. The pregnancy takes around four weeks. Normally around 50 to 300 baby seahorses are born in each brood; however, 581 babies were once documented in a single brood. The newborns are already around 0.6 inches long, whereas the adults grow up to six inches long.

It is difficult to keep this species in an aquarium, because long-snouted seahorses usually only accept live feed.

Linnaeus's Parade Seahorse

SHORT-SNOUTED SEAHORSES (*H. hippocampus*) are primarily found in the same European regions as long-snouted seahorses. They also grow to a length of up to six inches. As their colloquial name suggests, they have a shorter snout than most species. Their snout is often only one-third the length of their head. In addition, short-snouted seahorses do not have a conspicuous spiny mane of skin filaments. However, many specimens do have highly visible, rounded spines above their eyes. A special characteristic of this species is that the females "pale" at the culmination of their courtship to indicate their willingness to mate.

The short-snouted seahorse was scientifically classified in 1758 as the first horse of the sea by the founder of modern taxonomy himself, Carl Linnaeus. In contrast to most other seahorses, they prefer regions of the seafloor that are only sparsely vegetated, and they also live at greater depths in the sea.

Short-snouted seahorses are often dried and used as remedies in traditional Chinese medicine (TCM), despite existing laws that have prohibited fishing and trading them in most countries for almost twenty years (see chapter 13). Unfortunately, a black market has been established. While many other species of seahorse are able to adapt their coloring to camouflage themselves in their habitat, short-snouted seahorses cannot perform this trick—definitely a handicap in their fight for survival (see chapter 15).

Miniature Versions

DWARF SEAHORSES (*H. zosterae*), only grow to a length of about one inch. In contrast to the group of pygmy seahorses, which are also tiny but have fewer bony plates, these creatures are just like miniature editions of the larger species. Conspicuous characteristics are the rings on the trunk and the male's brood pouch, which is visible on the outside of his belly. This seahorse's coronet is a tall crown, which may look like a pillar or a doorknob. Depending on their habitat, they may be yellow, green, black, or with spots of various colors.

Dwarf seahorses live in shallow seagrass beds. They are rather common along the coast and brackish waters of Florida, but also can be found on the coastlines of Texas, Mexico, and the Bahamas. These little seahorses mainly feed on tiny copepods. Dwarf seahorses quickly achieve sexual maturity and can reproduce after just less than three months. Court-ship includes quivering, as well as rising to the water surface together, before the female deploys her ovipositor. The male pregnancy takes around twelve days, and some males mate again only a few hours after having given birth. Usually around twenty babies are born, yet sometimes up to fifty, during the mating season from February to October.

Dwarf seahorses are the slowest fish in the world, with a top speed of five feet per hour. However, their special stealth hunting technique makes them first-class predators (see chapter 6). These creatures are relatively good at adapting to changing environmental conditions. This is one of the rea-sons why small groups of dwarf seahorses are usually easy to keep in aquariums, where they may live up to two years. Usually, pairs consist of seahorses of a similar size.

Fiery Mustangs

LINED SEAHORSES (*H. erectus*) grow up to seven inches in length. They live in shallow coastal waters from Nova Scotia in Canada down the Atlantic coastline to Brazil. Depending on their immediate environment, lined seahorses' basic color varies from white to yellow, orange, red, or black. Regarding their size, body shape, and spine appearance, there is great variation among the different specimens of this species. This has resulted in recurring new "discoveries" by aficiona-dos, who propose them as new species. However, they do have two features that differentiate them clearly from all other horses of the sea: thicker bony trunk rings, as well as conspicuous spines on the first, third, and fifth trunk ring. The clearly visible lines on the trunk and the vertical lines on the neck are characteristic of this species.

It is still unclear whether the lined seahorses found on the coast of North America are just visitors from South America, or if a subpopulation has become endemic. Genetic sequencing of specimens caught near New York has provided evidence that these creatures are at home there, although they do leave the shallow coastal waters in the wintertime to find less turbulent regions of the Atlantic in greater depths.

Lined seahorses are popular pet fish and are advertised by some dealers under imaginative names, such as "sunburst," "mustang," or "sunfire." These creatures mate between May and October. Normally around 200 to 500 baby seahorses are born per brood, but there may be up to 1,300. They prefer habitats of seagrass, sponges, and drifting brown algae (*Sargassum*).

Real Freaks

The **PARADOXICAL SEAHORSE** (*H. paradoxus*), which was scientifically identified only a few years ago, can grow up to around 2.3 inches in length. Its head is bulbous, and it is laterally flattened. In addition, this particularly unusual horse of the sea has a row of flexible fin-like skin lobes down the middle of its trunk and tail.

It's not clear what color a live specimen of this spineless seahorse might be. Only a single one has ever been found—a dried, eggshell-colored specimen exhibited in the

South Australian Museum in Adelaide. During a geoscientific expedition in 1995, this marine creature was dredged up by a research ship off the southwest coast of Australia from a depth of 335 feet below sea level.

It spent fifteen years unnoticed in a box in the museum— until experts found it and classified this mysterious seahorse as a representative of a hitherto unknown species. It is also not clear how paradoxical seahorses propel themselves forward underwater, because they lack a dorsal fin—the organ that provides forward propulsion for all the other seahorses.

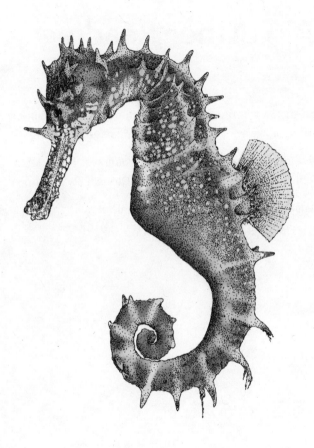

Spiny Seahorse
Hippocampus histrix

Further Reading

IMPORTANT SCIENTIFIC DETAILS for individual chapters of this book were found in the following reference works:

Poseidon's Steed: The Story of Seahorses, From Myth to Reality by Helen Scales (New York: Gotham Books, 2009).

Seahorses: A Life-Size Guide to Every Species by Sara Lourie (Chicago: University of Chicago Press, 2016).

Special Thanks

DR. DANIEL ABED-NAVANDI, Dr. David Booth, Dr. Sarah Foster, Dr. Julia Frankenstein, Peter Gutwasser, Dr. David Harasti, Prof. William Holt, Claudia Jürgens, Pavel Klinckhamers, Prof. Friedrich Ladich, Dr. Adam Lim, Dr. Anna Lindholm, Dr. Sara Lourie, Prof. Axel Meyer, Dr. Olivia Roth, Dr. Helen Scales, Dr. Ulrike Schmidt, Dr. Ralf Schneider, Katja Scholtz, Prof. Peter Teske, Elena Theys, Prof. Amanda Vincent, and Judith Weber.

Index